THE LITERATURE OF PHOTOGRAPHY

THE LITERATURE OF PHOTOGRAPHY

THE CHEMISTRY OF LIGHT AND PHOTOGRAPHY

BY

Dr. HERMANN VOGEL

ARNO PRESS
A NEW YORK TIMES COMPANY
NEW YORK ★ 1973

Reprint Edition 1973 by Arno Press Inc.

Reprinted from a copy in
The University of Illinois Library

The Literature of Photography
ISBN for complete set: 0-405-04889-0
See last pages of this volume for titles.

Manufactured in the United States of America

Library of Congress Cataloging in Publication Data

Vogel, Hermann Wilhelm, 1834-1898.
The chemistry of light and photography.

(The Literature of photography)
Reprint of the 1875 ed., which was issued as v. 15 of The International scientific series.
Translation of Die chemischen Wirkungen des Lichts und die Photographie.
1. Photography. 2. Photochemistry. 3. Photographic chemistry. I. Title. II. Series. III. Series: The International scientific series (New York), v. 14.
TR145.V88 1973 770'.1'54135 72-9242
ISBN 0-405-04946-3

THE INTERNATIONAL SCIENTIFIC SERIES.

VOLUME XIV.

PLATE I. *Frontispiece.*

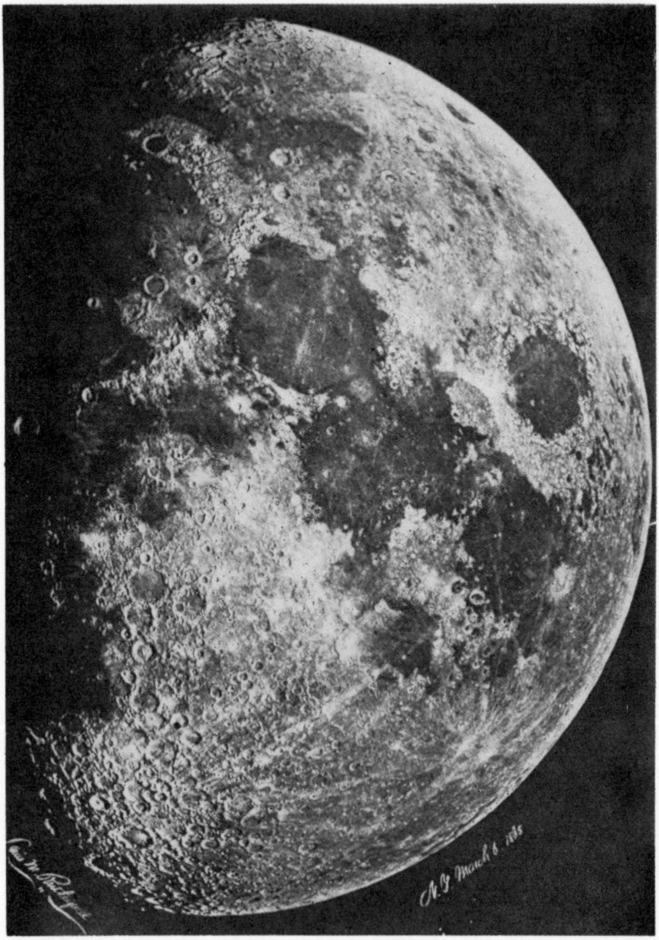

Woodbury process. *See p.* 193.

PHOTOGRAPH OF THE MOON,
(after RUTHERFORD'S original negative.)

THE INTERNATIONAL SCIENTIFIC SERIES.

THE CHEMISTRY OF LIGHT AND PHOTOGRAPHY.

BY

DR. HERMANN VOGEL,

PROFESSOR IN THE ROYAL INDUSTRIAL ACADEMY OF BERLIN.

WITH ONE HUNDRED ILLUSTRATIONS.

NEW YORK:
D. APPLETON AND COMPANY,
549 AND 551 BROADWAY.
1875.

PREFACE.

Among the splendid scientific inventions of this century, two are specially prominent—Photography and Spectrum Analysis. Both belong to the province of Optics, and at the same time of Chemistry. While Spectrum Analysis has, down to the present time, remained almost exclusively in the hands of the learned, Photography passed immediately into practical life, spread over almost every branch of human effort and knowledge, and now there is scarcely a single field in the universe of visible phenomena where its productive influence is not felt.

It brings before us faithful pictures of remote regions, of strange forms of stratification, of fauna, and of flora; it fixes the transient appearances of solar eclipses; it is of great utility to the astronomer and geographer; it registers the movements of the barometer and thermometer; it has found an alliance with porcelain painting, with lithography, metal and book typography; it makes the noblest works of art accessible to those of slender means. It may thus be compared to the art of printing, which confers the greatest benefit by multiplying the production of thought, for it conveys an analogous advantage by fixing and multiplying phenomena. But it does more than this. A new science has been called into being by Photography, the Chemistry of Light; it has given new conclusions respecting the operations of the vibrating ether of light. It is true that these services, rendered by Photography to art and science, are only appreciated by the few. Men of science have in great measure neglected this branch after the first enthusiasm excited by Daguerre's invention had evaporated; it is only cursorily that physical and chemical matters are treated on in manuals of Photography.

Taking this into consideration, it has seemed expedient to the Author to give a popular view of Photography and the Chemistry of Light, showing their important bearing on science, art, and industry. The Publisher has met the Author in the readiest manner, not only by providing numerous woodcuts to explain the text, but by obtaining specimens of the latest discoveries in Photography, at a considerable cost. So that as the tables annexed give a view of what is achieved by Photography in connection with typography, he trusts the work may meet a friendly reception.

THE AUTHOR.

BERLIN, *January*, 1874.

TABLE OF CONTENTS.

LIST OF ILLUSTRATIONS.

LIST OF PLATES.

THE
CHEMISTRY OF LIGHT.

CHAPTER I.

DEVELOPMENT OF OUR PHOTO-CHEMICAL KNOWLEDGE.

Diverse Kinds of Operations of Light—Physical and Chemical Changes—Bleaching Effect of Light—Effects of Light upon Chloride of Silver and Lunar Caustic (Nitrate of Silver)—Chemical Ink—Pictures upon Paper saturated with Nitrate of Silver—The Labours of Wedgwood and Davy—The Camera Obscura—Nièpce—Effects of Light upon Asphaltum—Heliography—Its Application to the Production of Paper Money—Iodide of Silver—Discovery of the Daguerreotype.

The light which radiates from the great central body of our planetary system produces manifold effects upon the animate and inanimate world, some of which are at once evident to the senses, and have been known for thousands of years, while others, again, are not so apparent to the eye, and have been discovered, examined, and utilized only through the observations of modern times.

The first effect which every person, however uncultivated, notices, when, after the darkness of night, the sun rises, is the visibility of bodies. The rays from the source of light are thrown back (reflected) from these different bodies, they reach our eyes, and produce an impression upon the retina, the result of which is the perception of material objects through the eye. But soon another effect is observed, not through the eye, but by sensation. The sun's rays not only illumine bodies upon which they fall, but heat them, as is felt when the hand is held in the rays. Both effects, the shining or illuminating, and the warming effects of the sunbeams, differ very essentially from each other. The illuminating effect we perceive instantaneously; the heating effect is only felt after a certain time, which may be shorter or longer, as the heating power of the sun is stronger or weaker.

In addition to these two effects of sunlight, there is a third, which generally requires a still longer period to make itself noticed, and which cannot be directly perceived either through the eye or by sensation, but only through the peculiar changes which light produces in the material world. These are the chemical effects of light.

If we take a small piece of wood, and bend or saw it, we change its form; if we rub it, it becomes warm; we change its temperature, but it still remains wood. These changes, which do not affect the substance or matter (*stoff*) of a body, we term physical.

But, if we set fire to a piece of wood, strong-smelling gases ascend, ashes are deposited, and a black residuum remains, which is totally different from the wood. By this process a quite different substance—charcoal—has been produced. *Material* changes of this kind we term chemical changes;—and such chemical changes are, in an especial manner, the result of heat. If, for instance, we heat a bright iron wire red hot, it undergoes apparently only a physical (not a material change). But, if we allow it to cool, we find the bright rod has become dull and black; that it has received a brittle, black surface, which, in the process of bending, easily breaks, and the bright, tough, flexible iron has undergone a chemical change. That is to say, a change of substance has taken place, the iron has been converted into another body, into *black oxide*, because it has combined with a component part of the surrounding air with oxygen.

Chemical changes of this kind are not only produced by heat, but also by light.

It has long been known that when the colours of which fabrics are dyed are what is called "not fast colours," they fade in the light, that is, become paler. In this case the coloured material changes into a colourless or differently coloured body; and that this is the effect of light is evident from the fact that that part of the material in question which is covered up from the light —that beneath the folds—remains unchanged. This discolouring effect of light has been long turned to practical use in the bleaching of linen. The unbleached fabric is spread out in the sunlight, and repeatedly moistened with water; and thus, through the combined effect of light and moisture, this dark colouring substance becomes gradually soluble, and can then be removed from the linen by boiling it in lie (alkali).

It was formerly believed that the changes we have just described were caused by the heat which is produced in bodies by the sun's rays. That this is an erroneous view is evident from the fact that fabrics dyed in colours which are "not fast" can be exposed for months together in the temperature of a hot oven without any bleaching effect; and further, that wax, which the sunlight likewise bleaches, becomes darker, rather than paler, through heat.

As we remarked before, the bleaching effect of sunlight is a slow process, and this circumstance renders the phenomenon less striking. A sudden and rapid occurrence surprises us, and stirs us up to inquire and to reflect.

In the mines of Friburg is now and then found a vitreous dull-shining silver ore, which, on account of its appearance, is called horn silver (muriate or chloride of silver).

This horn silver consists of silver and chlorine in chemical combination, and can be artificially produced by directing chloric gas upon metallic silver. This horn ore in its original position is completely colourless, but as soon as it is exposed to the daylight it assumes, in a few minutes, a violet tint. This effect of light has long excited the astonishment of men of science.

In another argentiferous substance this phenomenon is still more obvious. If nitric acid is poured upon silver, it dissolves with effervescence. If the liquid part of the solution be evaporated, a solid mass of crystals is obtained, which is not silver but a combination of silver with nitric acid. This nitric-acid silver is totally different from ordinary silver; it dissolves easily in water, like sugar; it has a bitter, disagreeable taste; it readily diminishes and destroys organic matter; and it is therefore used as a corrosive agent, under the name of lunar caustic, or nitrate of silver.

It has been long known that the fingers which grasp the lunar caustic, or anything that is sprinkled with a solution of it, quickly assume a dark colour. This can be at once tested by moistening a small piece of paper with the silver solution, allowing it to dry, and then placing it in the light.

These properties were soon made use of to produce a so-called indelible ink, which is nothing more than a solution of one part nitrate of silver in four parts of water, combined with a somewhat thick solution of gum. Written characters traced with it upon linen cloth are pale; but, when dried in the sunlight, quickly become dark brown, and are not injured by washing. Ink of this kind is much used in hospitals for marking linen. A quill, not a steel pen, must be made use of, as that metal decomposes the nitrate of silver. It is not unusual to print the characters by means of wooden type.

From the discovery of the blackening of paper saturated with lunar caustic to the invention of photography there was but a step; yet it was long before any one thought of producing pictures by the help of light alone, and still longer before these attempts were crowned with success.

Wedgwood, the son of the celebrated manufacturer of porcelain who produced the still popular Wedgwood ware, and Davy, the celebrated chemist, made the first attempts in the year 1802. They placed flat bodies, such as the leaves of plants, upon lunar caustic paper. Light was thus kept from the superimposed parts of the paper, the underlying parts remained white, whilst the uncovered portions of the paper were blackened by the light; and thus was produced a white outline, or "silhouette," of the superimposed objects upon a black ground. (See Figs. 1 and 2.)

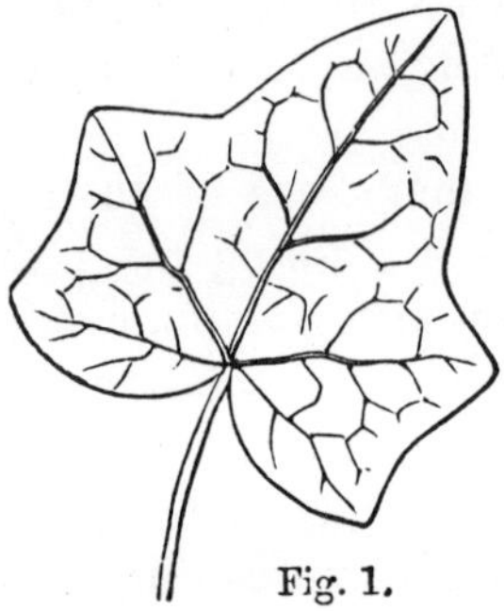

Fig. 1.

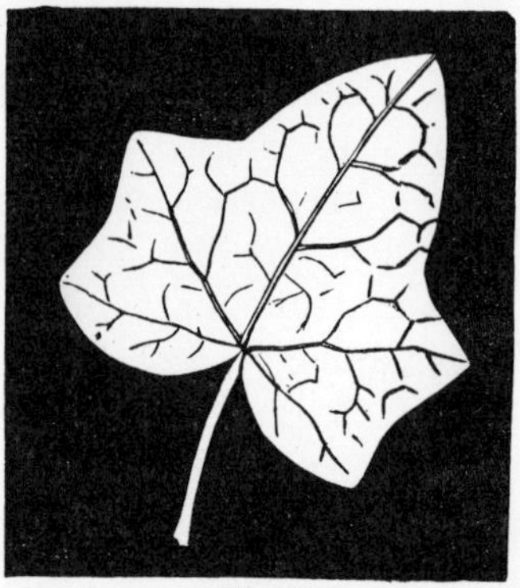

Fig. 2.

Wedgwood produced in this manner copies of drawings in white lines upon a black ground made upon glass, and this process became the basis, in modern times, of a mode of treatment which attained the highest importance, coming under the name of the light-*paus* process.

Unfortunately these pictures were not durable. They had to be kept in the dark, and could only be exhibited in a subdued light. If they remained long exposed to the light, the white parts also became black; and thus the picture disappeared. No means were then known to make the pictures durable, that is to say, to make them light-resisting, or as we now say, to *fix* them. But the first step towards the discovery of photography was made; and the idea of producing pictures of the world of matter without the help of the draughtsman became, after these first attempts, so extremely attractive that, from that time, both in England and in France, a large number of persons occupied themselves with the subject in private with the greatest enthusiasm.

It is clear that by the process of Wedgwood and Davy only flat bodies could be copied, and, notwithstanding all the improvements of which the process was still susceptible, it admitted of only a limited application.

But even Wedgwood seized the thought whether it were not possible by the help of light to produce pictures of any bodies whatsoever on sensitized paper. He tried to effect this by the aid of an interesting optical instrument which had the property of projecting flat-shaded images of solid objects. This instrument is the camera obscura.

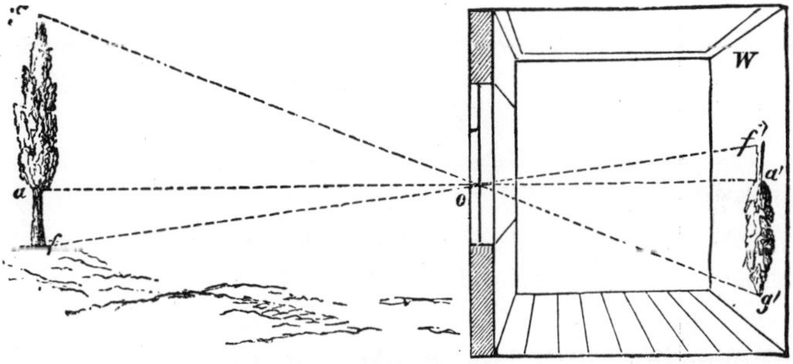

Fig. 3.

If a small hole be made in the window shutter of a completely darkened room on a sunny day, a clear image of the landscape will be seen on the opposite wall of the room.

Let *a* be a poplar, *o* the hole, and *w* the back wall of

the room, then from each point of the poplar rays of light are projected towards the hole, and beyond that in a straight line to the wall. It is now clear that to the point *a'* light can only arrive from the point *a* of the poplar, which is situated on the extension of the line *a'* and *o*. Therefore the point in question of the wall can only reflect light, which in its colour and position, corresponds to the point *a*. The same remark applies to the points *f* and *g*, and the result accordingly is that on the wall an inverted image of the tree is visible. This was first observed by Porta the celebrated Italian natural

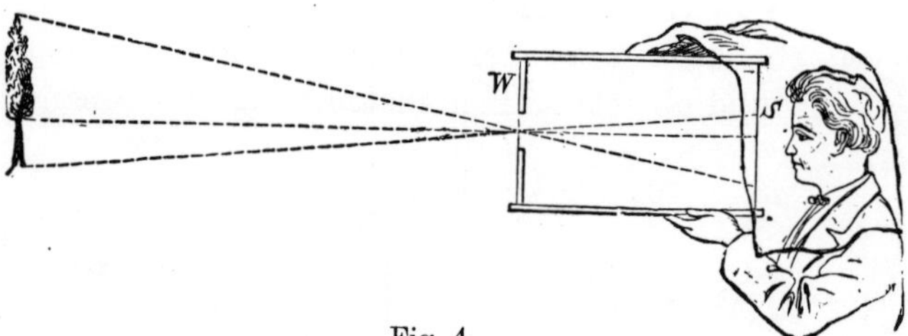

Fig. 4.

philosopher ot the 16th century, whose house, we are told by contemporaries, was seldom free from visitors in search of knowledge. This instrument was soon improved by substituting a small box (Fig. 4) instead of the chamber, which box in place of a solid wall had a movable unpolished slide. On this slide the image of an object in front of the box is clearly visible, if a minute hole is made in the front partition, which answers best if composed of a thin tin metal plate.*

These images appear still more beautiful if, instead of

* To prevent the access of light the head must be covered with a cloth.

a hole, a glass lens, or, as it is called, a focal lens, is substituted. This focal lens, or "burning-glass," at a certain distance, which is equal to that of its focus, projects a distinct image of the objects—which is much better defined and clearer than that which is produced by the hole.

In this improved form was the instrument now employed by Wedgwood and Davy. Their idea was to catch a small picture upon the unpolished slide by means of sensitized paper. They fastened a piece of paper saturated with salts of silver upon the place of the image, and left it there for several hours—unfortunately without result. The pictures were not distinct enough to make a visible impression upon the sensitized paper; or the paper was not sufficiently sensitive. It now became necessary to find a more sensitized preparation to catch and to retain the indistinct image; and this was achieved by a Frenchman—Nicéphore Nièpce. He had recourse to a very peculiar substance, the sensitiveness of which to light was before unknown to any one—asphaltum, or the bitumen of Judæa. This black mineral pitch, which is found near the Black Sea, the Dead Sea, the Caspian, and many other places, is soluble in ethereal oils—such as oil of turpentine, oil of lavender, besides petroleum, ether, and others. If a solution of this substance is poured upon a metal plate, and allowed to cover the surface, a thin fluid coating adheres to it, which soon dries and leaves behind a light brown film of asphaltum. This film of asphaltum does not receive a darker hue in the light, but it loses by light its property of solubility in ethereal oils.

If such a plate, therefore, is put in the place of the

small image of the camera obscura, the asphaltum coating will remain soluble on all the dark places (shadows) of the image, whilst on the light spots it will remain insoluble. The eye, it is true, does not perceive these changes. The plate appears the same after as before being exposed to the influence of light. But, if oil of lavender is poured over the coating of asphaltum it dissolves all the spots that had remained unchanged, and leaves behind all those that had been changed by light, that is, had been rendered insoluble. Thus, after several hours exposure in the camera obscura, and subsequent treatment with ethereal oils, Nièpce succeeded, in fact, in obtaining a picture. This picture was very imperfect it is true, but still interesting as a first attempt to fix the images of the camera, and still more interesting as evidence that there are bodies which lose their solubility in the sunlight. These facts were again verified long after the death of Nièpce, and they led to one of the finest applications of photography, that of heliography, or the combination of photography with copper-plate printing, which Nièpce himself to all appearance had already known.

A copper-plate print is produced in this way:—A smooth copper plate is engraved with the burin (or graving tool), that is to say, the lines which should appear black in the picture are deeply incised in the plate. In producing impressions, ink is first rubbed into these incisions, and then a sheet of paper is placed upon the plate and subjected to the action of a cylindrical press; after which the ink passes over the paper and produces the copper-plate impression. Nièpce endeavoured to substitute, by the help of light, a less laborious process than the troublesome one of cutting by the engraver. To effect this he covered the copper plate with asphaltum as before stated, and exposed this to the light beneath a drawing on paper. In this case the black lines of the drawing kept back the light; and, accordingly, in these places the asphaltum coating remained soluble; under the white paper, on the contrary, it became insoluble. Therefore, when lavender oil was afterwards poured over the plate, the parts of the asphaltum which had become insoluble adhered to the plate, whilst the soluble parts were dissolved and removed; and thus the plate in those places was laid bare. Thus a film of asphaltum is obtained on the plate in which the original drawing appeared as if engraved.

If a corrosive acid is now poured on such a plate, it can only act on the metal in those places where it is not protected by the asphaltum; and in these places the metal plate was in fact eaten into. Thus an incised drawing upon a metal was effected through the corrosive operations of the acid, and a plate was obtained which, when rubbed clean, could be used for impressions like an engraved copper plate. Copper-plate impressions of this kind have been found amongst the papers left behind by Nièpce, which he called "heliographs," and showed to his friends as far back as 1826. This method, in an improved form, is still in use at the present day, especially in the printing of paper money, when it is requisite to produce a number of engraved plates which are all to be absolutely alike, so that one piece of paper money may perfectly correspond to another, and may therefore be distinguished from counterfeits. In this way the arms and the inscription on the upper side of the Prussian ten thaler notes are printed off from heliographic plates. Thousands of people carry photographic impressions in their pocket-books, without knowing it. Nor is there any occasion to fear that these notes can be imitated easily by the help of photography or heliography. We shall show later on, that the ground tint, the paper itself, and the colour of the inscription present well-devised obstacles to all such imitations, and make them very difficult, if not impossible.

Nièpce's impressions were undoubtedly very imperfect, and therefore remained unnoticed. He himself gave them up, and again entered upon a series of experiments to fix the charming images of the camera obscura. In 1829 Daguerre joined him; and both carried on experiments in common until 1833, when Nièpce died without having obtained the reward of his long-continued efforts.

Daguerre went on with his experiments; and he would not, perhaps, have carried them any further if a fortunate accident had not worked in his favour.

He made experiments with iodide of silver plates. Then he produced, by exposing silver plates to the vapour of iodine, a peculiar and very volatile chemical element. Under this treatment, the silver plate assumed a pale yellow colour, which is peculiar to the combination of iodine and silver. These plates of iodide of silver are sensitive to light, they take a brown colour when exposed to it, and an image is soon produced upon them when they are exposed to the operations of light in the camera. A very long exposure to light, however, is necessary to this end; and the thought could scarcely have arisen of taking the likeness of any person in this manner, for he would have been obliged to remain motionless for hours to obtain it.

One day Daguerre placed aside as useless, in a closet in which were some chemical substances, several plates that had been exposed too short a time to the light, and therefore as yet showed no image. After some time he looked by accident at the plates, and was not a little astonished to see an image upon them. He immediately divined that this must have arisen through the operation on the plates of some chemical substance which was lying in the closet. He, therefore, proceeded to take one chemical out of the closet after the other, and placed in it plates recently exposed to the light, when, after remaining there some hours, images were again produced upon them. At length he had removed in succession all the chemical substances from the closet; and still images were produced upon the plates that had been exposed to the light. He was now on the point of believing the closet to be bewitched, when he discovered on the floor a shell containing quicksilver, which he had hitherto overlooked. He conceived the notion that the vapour from this substance—for mercury gives off vapour even at an ordinary temperature—must have been the magic power which produced the image. To test the accuracy of this supposition, he again took a plate that had been exposed to light for a short time in the camera-obscura, and on which no image was yet visible. He exposed this plate to the vapour of quicksilver, and, to his intense delight, an image appeared, and the world was again enriched by one of its most beautiful discoveries.

CHAPTER II.

THE DAGUERREOTYPE.

Its Publication and Spread—Its Path of Development—Improvements—Discovery of the Portrait Lens—Æsthetic effects of the Daguerreotype.

MANY persons at the present day who have before their eyes the grand productions of paper photography, such, for example, as portraits of life-size, view doubtless with pity, or even contempt, the little pictures that were called daguerreotypes from their inventor. The appearance of these pictures was no doubt injured by the ugly mirror-like dazzle which prevented a clear view of them. It was different in the year 1830, when Daguerre's discovery was first spread abroad by report. Pictures were said to be produced without a draughtsman by the operations of the sun's rays alone. That was of itself wonderful; but it was still more wonderful that, by the mysterious operation of light, every body impressed its own image on the plate. How many extravagant hopes and how many evil prognostications were associated with the report of this mysterious invention.

It was prophesied that painting would come to an end, and that artists would die of starvation. Every one hoped that he could with ease obtain images of any objects which he desired.

A friend is leaving home: in an instant his image is permanently retained at the moment of departure. A joyous company is assembled: a picture is taken of it at once as a souvenir. All objects were thus retained as pictures by the chemical action of the rays of light: the landscape glowing with the magical effects of sunset, the favourite spot in the garden, the daily motley movement of the streets, of men, of animals, of everything alive.

Then came sceptics who declared the whole thing impossible. These persons were reduced to silence by the testimony of Humboldt, Biot, and Arago, the three celebrated natural philosophers to whom Daguerre disclosed his secret in 1838. The excitement grew. Through the influence of Arago an application was made to secure to Daguerre a yearly pension of 6000 francs, provided he made public his discovery. The French Chamber of Deputies agreed; and, after a long and tiresome delay, the discovery was at length disclosed to the expectant world.

It was at a memorable public *séance* of the French Academy of Sciences in the Palais Mazarin, on the 19th of August, 1830, that Daguerre, in the presence of all the great authorities in art, science, and diplomacy, who were then in Paris, illustrated his processes by experiment.

Arago declared that "France had adopted this discovery, and was proud to hand it as a present to the whole world;" and henceforth, unhindered by the quackery of mystery, and unlimited by the right of patent,* the discovery of Daguerre made the round of the civilized world.

Daguerre quickly gathered round him a number of disciples from all quarters of the globe; and they transplanted the process to their homes, and became in their turn centres of activity, which daily added to the number of disciples of the art.

Sachse, a dealer in art still living in Berlin, was initiated into Daguerre's discovery on the 2nd of April, 1839, and was appointed Daguerre's agent in Germany on the 22nd of September, four weeks after the publication of the discovery. Sachse had already produced the first picture at Berlin. These pictures were gazed at as wonders, and each copy was paid for at the rate of from £1 to £2; while original impressions of Daguerre fetched as much as £4 16*s*. 8*d*. (120 francs). On the 30th of September Sachse made experiments in the Park of Charlottenburg, in the presence of King Frederick William the Fourth. In October the earliest Daguerre apparatuses were sold in Berlin. The first set of apparatus was purchased by Beuth for the Royal Academy of Industry at Berlin; and it is still to be seen there. After the introduction of the apparatus, it was in the power of every one to carry out the system; and a great number of daguerreotypists started into existence. Men of science, too, cultivated (more than they do now) the new art: among others, the natural philosophers, Professors Karsten, Moser, Nörrenberg, Von Ettinghausen—nay, even ladies, as Frau Professor Mitscherlich. The first objects photographed by Sachse

* It was only in England that the process was patented, before its publication on the 15th of July, 1839.

were architectural views, statuary, and paintings, which for two years found a ready sale as curiosities. It was in 1840 that he first represented groups of living persons, and in this way photography became especially an art of portraiture. It made the taking of portraits its principle means of support, and in two years there were daguerreotypists in all the capitals of Europe.

In America a painter, Professor Morse, afterwards the inventor of the Morse telegraph, was the first to prepare daguerreotypes; and his coadjutor was Professor Draper.

Let us now consider more closely the process employed in producing daguerreotype plates. A silver plate, as I have said, or in the place of it a silver-plated copper plate, serves as a plane surface for the image. This is rubbed smooth by means of tripoli and olive oil; and then it receives its highest polish with rouge and water and cotton. It is only a plate so extremely well polished that can be used for the process. This burnished plate is placed with its polished side upon an open square box, the floor of which is strewn with a thin layer of *iodine*. This *iodine* evaporates, its vapours come into contact with the silver, and instantly combine with it. By this means the plate first assumes a yellow straw-colour, next red, then violet, and lastly blue. The plate is then protected from the light; next it is placed in the camera obscura, where the image on the ground-glass slide is visible, and "exposed" for a certain time. It is afterwards brought back into the dark, and put into a second box, upon the metal floor of which there is quicksilver. This quicksilver is slightly warmed by means of a spirit lamp. At first not a trace of the image is visible on the plate. This first comes out when the vapour of the quicksilver precipitates itself upon the places affected by the light, and the result is in proportion to the effectiveness of the operation of light. During this process the quicksilver condenses itself into very minute white globules, which can be very well discerned under the microscope. This operation is called the development of the picture.

After the development the iodide of silver, being sensitive to light, must be removed to render the image durable, that is, "to fix it." This is effected by using a solution of *hypo-sulphite of soda,* which dissolves the *iodide of silver.* Nothing more is required after this than to wash with water and dry, and the daguerreotype is completed. Sometimes, in order to protect the picture, it was usual to gild it. A solution of chloride of gold was poured over, and then it was warmed; a thin film of gold was deposited, which contributed essentially to the durability of the pictures. A picture of this nature, however, remains always exposed to injury, and requires the protection of frame and glass.

Daguerre's first pictures needed an exposure of 20 minutes to the light—too long for taking portraits. But soon after it was found that bromine, a rare substance having many points of resemblance with iodine, employed in combination with the latter, produces much more sensitive plates, which required far less time, perhaps not more than from one to two minutes, for exposure.

Many of us, perhaps, still remember the early period of photography, when persons were obliged to sit in the full sunlight, and allow the dazzling rays to fall directly upon the face—a torture which is clearly marked and visible on the portraits still preserved of these photographic victims, in the blackened shadows, the distorted muscles, and the half-closed eyes. These caricatures could certainly not bear any comparison with a good portrait from the life, nor probably would portrait-photography have ever had such success if it had not succeeded in obtaining the exposure to a moderated light. This was obtained by the invention of a new lens—the double objective portrait lens of Professor Petzral, of Vienna.

This new lens was distinguished by the fact that it produced a much clearer picture than the old lens of Daguerre, because it was now possible to take impressions from less dazzlingly lighted objects. This lens was invented by Petzral in 1841. Voigtländer ground the lens according to his directions, and soon one of Voigtländer's lenses became indispensable to every daguerreotypist. By employing *iodide of bromium* and Voigtländer's lens, the process of exposure was made a matter of seconds.

The daguerreotype art had thus reached its zenith. However delicate pictures produced appeared, it was found, after the first enthusiasm had gone, and had given place to a cold spirit of criticism, that much still remained to be desired.

First, the gloss and brilliancy of the pictures make it difficult to look at them. Then there are several marked deviations from nature: yellow objects often produced little or no effect, or gave a black impression; on the other hand, blue objects, which appear dark to the eye, frequently, though not always, came out white.

This is still the case in photography, only now the

attempt is made to diminish this defect by subsequent treatment of the plate (negative re-touching).

But still a well-grounded æsthetic objection was brought against these pictures.

It was indisputable that the daguerreotype greatly surpassed painting by the wonderful clearness of detail, by the fabulous truthfulness with which it reproduced the outlines of objects. The daguerreotype plate gives more than the artist, but for that very reason it gives too much. It reproduces the subordinate objects as faithfully as the principal object in the picture.

Let us take the simplest case—a portrait. A painter, when he paints a portrait, does not by any means paint all that he sees in nature. The original wears, perhaps, a shabby coat, which shows a good many creases, perhaps a spot of grease, or a patch; but this does not distress the painter in the least, for he leaves out these accidental details. In the same spirit, if the original is seated before a whitewashed wall, the artist by no means puts a whitewashed wall into his picture, for he can leave out all that is displeasing, or add, on the contrary, what he wishes.

It is different in photography. This art, in taking portraits, reproduces all those minor accessories which disturb the picture, as faithfully as the principal object in it—the individual himself. Another point must be added to this. The different elements admitted by the painter into his picture are by no means made equally prominent. The head is the chief consideration in every portrait. The painter accordingly gives his best skill and care to the painting of the head in the most careful manner. At the very least he puts the head in the strongest light, and leaves the rest of the picture in a half-shade. But in photography it is by no means the head which is generally the most prominent—frequently it is a chair, or part of a background; and this detracts considerably from the effect of the picture. Finally, the expression of the face is an important point in a picture; and this varies with the mood of the sitter. Photography gives the expression which the original had at the moment the picture was taken. Now the expression varies, and is affected by a slight annoyance, a vexatious circumstance, *ennui*, or even by the motionless attitude which has to be observed during the process; and hence the portrait often looks strange and unnatural.

It is quite otherwise with painting. The painter has longer sittings of the original than the photographer; he soon learns to distinguish the accidental frame of mind from the characteristic expression of the face, and thus he is in a condition to produce a portrait much more closely corresponding with the character of the original than that of the photographer can ever be. This naturally applies only to paintings of masters of the first order. In the portraits of the dauber, none of these advantages are found; and this large class disappeared, like bats before the light, when the art of sun-painting suddenly rose upon the world. Many of these themselves adopted the new art and attained to greater results than they could have done as painters.

The artist of merit has no cause to fear photography. On the contrary, it proves advantageous to him by the fabulous fidelity of its drawing—through it he learns to reproduce the outline of things correctly—nor can it be disputed that, since the invention of photography, a decidedly greater study of nature and a greater truthfulness are visible in the works of our ablest painters.

We shall see further on, how even photography appropriated the æsthetic principles according to which painters proceed in preparing their portraits, and how thereby a certain artistic stamp was given to these productions, which raised them far above those of the early period. But this result was only possible when the technical part of photography had been brought to perfection, and a material better adapted to artistic work than an unyielding silver plate had been introduced.

CHAPTER III.

PAPER PHOTOGRAPHY AND THE *LICHT-PAUS*, OR NEW TALBOT PROCESS.

Talbot's Paper Photographs—*Licht-Paus* Paper—Leaf-prints—*Licht-Paus* Process and its Application.

In the same year that Daguerre published his process for the production of images on silver plates, Fox Talbot gave to the world a process for preparing drawings on paper by the help of light. Talbot was an English gentleman of fortune, who, like many Englishmen of leisure and means, employed his time in scientific observations. He plunged paper into a solution of kitchen salt, dried it, and then put it into a solution of silver. In this way he obtained a paper which was much more sensitive to light than that employed by Wedgwood. He employed this paper in copying the leaves of plants. Talbot himself says, "Nothing gives more beautiful copies of leaves, flowers, etc., than this paper, especially under the summer sun; the light works through the leaves, and copies even the minutest veins."

This is no exaggeration of Talbot. In the hands of the author there are impressions of this kind, from Talbot's own hand, which show excellently well the venous structure.

The pictures copied in this way in the sunlight are

naturally not durable, because the paper, by having salts of silver in its composition, is still sensitive to light. But Talbot offered the means of fixing the pictures—he plunged them in a hot solution of kitchen salt; in this way the greater part of the salts of silver was removed, and the pictures did not become obscure to any considerable extent in the light.

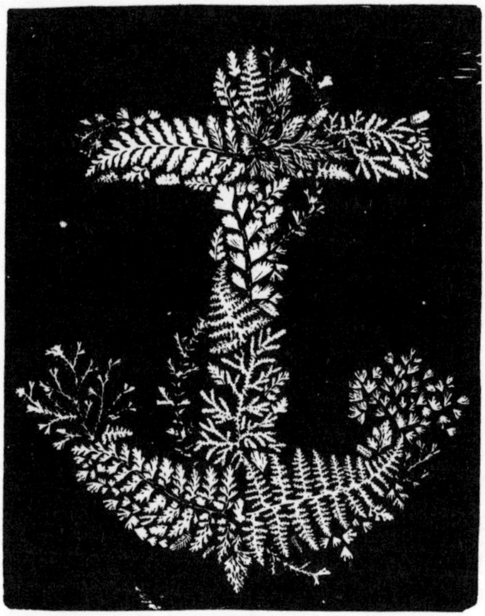

Fig. 5.

The celebrated Sir John Herschel carried out this fixing process even more successfully by plunging the pictures into a solution of hypo-sulphite of soda. This salt, which dissolves all the salts of silver, was at that time very expensive, costing 6 shillings per pound. The production of this salt soon kept pace with the increasing demands of photography, and now it is offered for sale by the ton, and at as low a rate as $6\frac{1}{4}d.$ the pound.

By this means the production of a durable sun-picture on paper, which Wedgwood had in vain attempted, was rendered possible. This method gave, no doubt, only pictures of flat objects which could be easily pressed on paper; for instance, leaves of plants, patterns of stuffs. The process has lately been resumed, after it had almost been forgotten. Charming ornaments of leaves, different plants, and flowers were produced; and these copies were proportionally more beautiful than the earlier ones, because a much finer and even-surfaced paper than that of Mr. Talbot has recently passed into trade, under the name of *licht-paus papier.** These prints are much liked in America. We give on the accompanying page a faithful imitation of one of these leaf-prints.

Since, by the cheapness of sensitive paper, the production of these leaf-prints has been made very easy, we give here the mode of producing them for our fair readers, who will be able in this manner, like their sisters in America, to make ornamental pictures for the adornment of lamp shades, portfolios, and similar things.

The leaves—especially ferns and the like—are suitably chosen, pressed between blotting-paper and dried, then gummed on one side and grouped gracefully by the fair artist upon a glass slab or plate, in a small frame (Fig. 6). As soon as the whole is dry, the impression can be at once commenced.†

* This paper is produced by Mr. Romain Talbot, 11, Karlstrasse, Berlin.

† These wooden frames, called copying frames, dishes, and fixing salt, are also manufactured by Mr. Talbot, at Berlin. There is now even a small plaything of this kind on sale, known by the name of the "sun-copying machine."

A small piece of sensitized paper is placed on the leaves after they are arranged, the two wooden lids, *h h*, are laid upon it, and fastened down by means of two little cross-bar pieces of wood, *x x*, and then the whole is exposed to the light, the glass side uppermost. The sheet of paper very soon assumes a brown colour, where it is not covered by the leaves, and

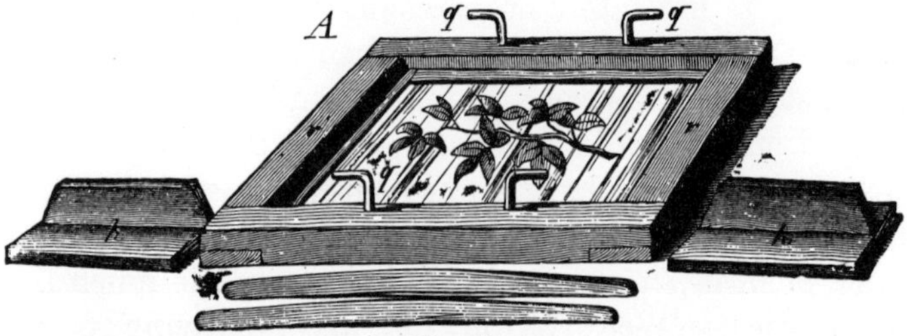

Fig. 6.

ultimately it receives a decided bronze tint. The light also penetrates partially through the leaves, and colours the paper lying under them brown. It is easy to discern how far the colouring has passed under the leaves, if one of the cross-bars, *x*, and the half cover, *h*, are removed, and the paper is lifted up.

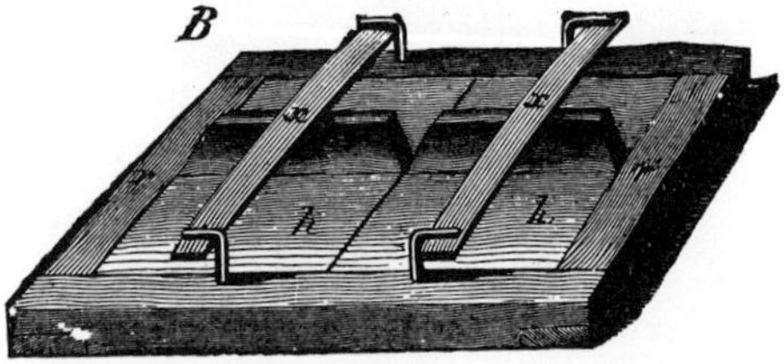

Fig. 7.

As soon as the impression is dark enough—it is quite a matter of taste whether the shade be dark or light—the paper is taken out and placed for a time in a dark closet. Several pictures can, in like manner, be taken one after the other, and these can be afterwards fixed, that is, made permanent in the light.

To this end the picture is placed in a flat dish (Fig. 8) containing water, for about five minutes, and then in a second dish in which a solution of twenty grammes of fixing natrium have been combined with a hundred grammes of water. The moment the impression is dipped in this it becomes of a yellowish brown. After the impression has remained ten minutes in the fixing solution—several leaves can be immersed in succession —it is taken out and placed in fresh water (most conveniently in a saucer). This operation of placing in fresh water is repeated from four to six times, the picture being left in the water three minutes each time.

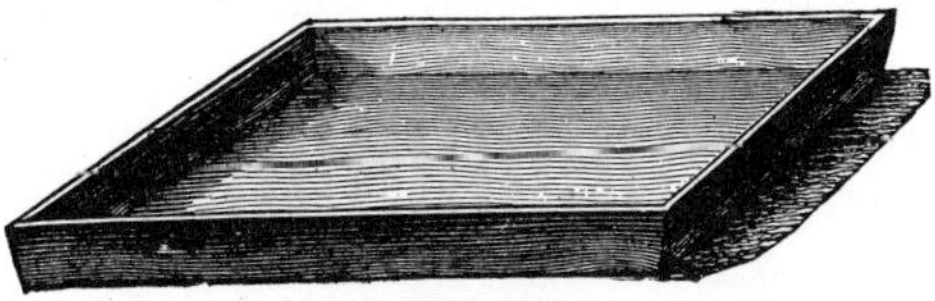

Fig. 8.

Afterwards the pictures are placed on blotting paper and suffered to dry; they can then be pasted upon cardboard, thick paper, linen, glass, or wood.

To many persons this process will appear an agreeable pastime, but latterly it has gained increasing consideration as an aid to the copying of drawings, maps, plans, copper-plate impressions, and so forth.

This work of copying, which used to cost the artisan and artist many hours of time and labour, and yet was inaccurate, can be accomplished with the least possible trouble by the help of the process described above.

Let the reader imagine a drawing placed on a piece of sensitized paper, and, after being firmly pressed together by a glass slab, exposed to the light. The light penetrates through all the white places of the paper, and colours brown those parts of the paper lying under them; but the black lines of the drawing keep back the light, and thus the underlying paper becomes white in these places. Therefore, if sufficient time is given for the operation of the light, a white copy on a dark brown ground is obtained in this manner, which is fixed and washed exactly like the leaf-prints described above. This copy is reversed with reference to the original, like an object and its reflection in a mirror; in other respects it is a faithful representation, stroke for stroke.

We give in Plate II. the copy of a woodcut struck off according to this method. This copy is rather too small, but the largest as well as the smallest drawing can be copied equally well; and copies of this kind, from drawings of the size of 4·11 feet, are made in technical offices, in mines, and in the manufactories of machinery.

Large copper frames are used for this purpose, resembling in their construction the small frames named above; and large wooden dishes, covered with a coating of asphaltum, are employed for fixing and moistening. This operation is called in practice *licht-paus* process. The black copy produced by it is called a negative picture, but a second white copy can be prepared from this by placing the negative upon sensitized paper; then the light shines through all the clear lines, and colours the paper lying under them of a dark hue, whilst it remains white under the dark places of the negative.

In this manner a picture is produced which perfectly resembles the primary original, called a positive. The washing and fixing are carried out just as scrupulously with the positive as with the negative. Fig. 9 offers a positive of this kind taken from the negative, Fig. 5.

Thus the geographer is in a position to prepare quickly faithful copies of his sketches and maps, the engineer

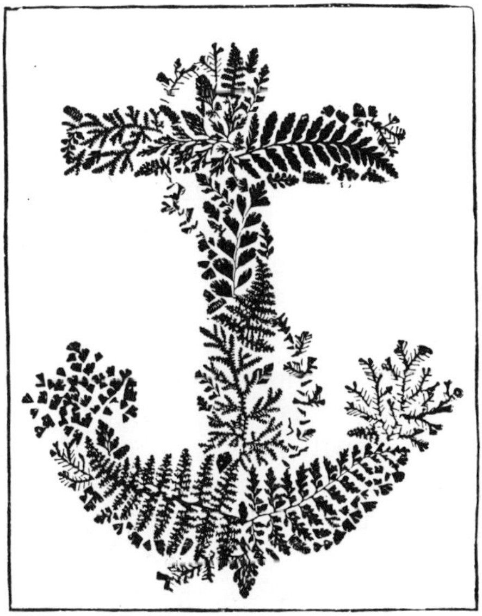

Fig. 9.

is able to copy the drawings of machines which are to serve as models for the workmen, and the student can copy illustrations of natural history which are to assist him in his studies. In the process of copying, the sensitized paper—*licht-paus* paper—must closely touch the original picture; therefore the former must be placed on the side of the picture, and not on the reverse side of the original.

This process has already done good service in military operations, where it was important to make quickly a copy of some map of which there was only one impression. If an attempt had been made to draw a copy of the map, it would have required several days to carry out, nor would the copy have been as correct as the *licht-paus*.

It is remarkable that this process, so important for industry, has only quite recently been known in its full value, although the experiments of Talbot have been before the world for thirty years. The explanation of this fact is, no doubt, to be found in the circumstance that the paper impressions were far less distinct than now, being often rendered worthless by spots. Another reason has been that the preparation of the paper requires especial care, and therefore frequently failed in the hands of the inexperienced; that is, of those who were not professional photographers. Further, the papers prepared according to the old method soon spoil, and had on that account to be used immediately after they were prepared.

These disadvantages have been removed by the invention of Romain Talbot's *licht-paus* paper, which is sold ready prepared, and can be kept for months; and by this means the process can be easily made available by every professional man and amateur.

CHAPTER IV.

THE DEVELOPMENT OF MODERN PHOTOGRAPHY.

Talbot's Paper Negatives—Photography as an Art for Multiplying Copies—Services of Nièpce de St. Victor—White of Egg Negatives—Gun-Cotton in Photography—Collodion—Archer's Negative Process—White of Egg Paper—Carte de Visite—Photographic Album.

THE reader has already learnt in the previous chapter what constitutes a negative, and how by its means copies produced by light, of plane objects, can be obtained.

Talbot, the inventor of this paper process, carried out further researches, in order to represent on paper, by the help of the camera-obscura, material objects which cannot be pressed together with sensitized paper; for example, a person or a landscape.

He attained this object two years after Daguerre's discovery, by means of paper prepared with iodide of silver.

He saturated paper in a solution of nitrate of silver, and then in a solution of iodide of potassium. He thus obtained a slightly sensitized paper, but one that could always be rendered very sensitive, by plunging it into pyrogallic acid and silver (gallate of silver).*

* The nature of this peculiar process is explained further on.

When this paper was exposed to the light in the camera obscura, it did not give at once a picture—this was only clearly defined after lying some time in the dark, or by subsequent treatment with pyrogallic acid and silver—but it came out as a negative, and not as a positive. Thus in taking, for example, a portrait, the shirt appeared black, also the face; while the coat, on the contrary, came out white.

The picture was made light-resisting by plunging it in a solution of hypo-sulphite of soda.

A negative thus obtained is a picture on a plane surface of a solid object, and Talbot prepared positive pictures from negatives of this kind.

He placed the negative upon a piece of sensitive paper saturated with chloride of silver, as described in the last chapter, and let the light work upon it. This shone through the white places of the negative, and imparted a dark colour to those parts of the sensitive paper lying under them, while the dark places of the negative protected the paper lying under them from the effects of the light. Thus he obtained a positive picture from a negative. He was now able to repeat the process as often as he pleased, and therefore was in a position to copy, by means of the light, many positives from a single negative. Photography was thus classed among the arts that repeat copies, and this circumstance exercised an important result on its future development.

Daguerre's method only gave a single positive at a time; if more were required, the person had to sit several times. In Talbot's method a single *séance* sufficed to produce hundreds of pictures.

It must be admitted that the earlier pictures of the

Talbot process were not remarkably engaging. Every roughness of the paper and each small speck of dirt were imprinted on the positive, which could not be compared in point of delicacy with the fine daguerreotypes; but the method was soon improved.

Nièpce de St. Victor, nephew of Nicophore Nièpce, the friend of Daguerre, had the happy idea of substituting glass for paper. He covered glass plates with a solution of white of egg, in which iodide of potassium was dissolved.

A solution of this kind can be easily produced by beating up the white of egg to the consistency of snow, and allowing it to deposit. The glass plates, after being dried and covered with a coating of white of egg, were afterwards dipped in a solution of silver. Iodide of silver was formed in this manner—the whole coating of the white of egg was coloured yellow, and became very sensitive to light.

Nièpce put these glass plates into the place of the picture in the camera-obscura, and suffered the light to work upon it.

Its impression was at first invisible, but afterwards became clearly perceptible when the picture was immersed in a solution of pyrogallic acid. Thus Nièpce obtained a negative on glass without the blemishes which appeared on paper negatives.

He repeated this negative exactly according to the same process that had been employed by Fox Talbot, and he obtained from the fine negative a correspondingly fine positive, which was much better calculated to bear a comparison with the productions of Daguerre.

Nièpce invented his method in 1847. It excited much

attention, but had a shady side: the treatment with white of egg, salts of silver, and pyrogallic acid was a dirty process. Therefore the method appeared to many, who had been accustomed to the daguerreotype, dirty and unpleasant, and deterred persons from trying it.

On the other hand, the advantages of the new process in repeating impressions was so evident that it could not be overlooked; therefore, even those who had an antipathy to soil their fingers nevertheless zealously devoted themselves to the work.

The easy decomposition of white of egg was, however, a great disadvantage in the new process. They sought to avoid this by adopting a more durable substance.

This was afforded by a new discovery, gun-cotton, made by Schönbein and Böttcher in 1847. Schönbein found that ordinary cotton saturated in a mixture of nitric acid and sulphuric acid assumes explosive properties similar to those of gunpowder. It was conceived that this substance was an important substitute for gunpowder, but it was soon found that its explosive property was very unequal, being sometimes too strong and at other times too weak. On the other hand, another very useful property was observed in the same body—that of being dissolved in a mixture of alcohol and ether. This solution leaves behind it a transparent membrane forming an excellent sticking-plaster for wounds. Thus the same substance that was destined to be a substitute for gunpowder, as a destructive agent for *producing* wounds, became actually a *remedy* for the latter. This solution of gun-cotton was called collodion.

The thought occurred to different photographic experimenters to try this substance instead of the white of egg, by coating glass plates with it; but the attempts did not at first lead to any satisfactory results. At length Archer published in England a full description of a collodion negative process surpassing in the beauty of results, in simplicity and security, Nièpce's white of egg process.

Archer coated glass plates with collodion, in which salts of iodide had been dissolved; he plunged this in a solution of silver, and thus obtained a membrane of collodion saturated with sensitive iodide of silver, which he then exposed in the camera.

The invisible impression of the light thus produced became visible by pouring gallic acid over the plate, or the still more powerful chemical agent, pyrogallic acid; or, instead of this, a solution of green vitriol.

A very delicate, clear negative was directly obtained by this process, which yielded much more beautiful impressions on paper than the original negative paper of Talbot. A very essential improvement was subsequently made in the preparation of negative paper by coating it with white of egg, according to the process of Nièpce de St. Victor. By this means it received a brilliant surface, and when exposed to the light it took a more beautiful and warmer tone, which gave the pictures a brighter appearance than those produced upon the ordinary paper.

Thus the Talbot-type, which at first seemed hardly worth notice compared with the process of Daguerre, was gradually so perfected by successive improvements, that it ultimately took precedence of Daguerre's. After 1853, paper pictures on collodion negatives came more and more into vogue, the demands for daguerreotypes fell off and soon vanished altogether, and were produced only here and there in America.

The collodion process is now the one universally employed. It acquired an immense impetus through the introduction of cartes de visite. These small portraits, which are intended to be given away, and therefore had to be produced in large numbers, were invented by Disderi, the court photographer of the Emperor Napoleon, and obtained so great a success that they were immediately introduced into all circles, and soon became a necessity for everybody. The moderate price at which these portraits were sold made them attractive to the smallest purses, and the general public crowded to the ateliers, the number of which increased daily.

The old-fashioned album, the favourite souvenir of young people, was now superseded by the carte de visite, and the portraits of friends were substituted for their written words. A photographic album is now found in every home; and in Berlin alone there are at present more than ten photographic album manufactories, to satisfy the demand, from whence they are exported to all parts of the world.

CHAPTER V.

THE NEGATIVE PROCESS.

The Dark Chamber—Light Inoperative Chemically—Plate-cleaning—Application of Collodion—Sensitizing—The Camera—The Arrangement—The Exposure to Light—The Developing Process—The Fortifying Process—The Fixing Process—The Varnishing.

In the previous chapters we have dwelt on the development of this art, and we are now able to feel at home in the atelier of a photographer. His whole business depends on the chemical operation of light,—and yet the scene of his principal activity is not the illuminated atelier, but a dungeon, in which the deepest night prevails, and which is called the dark chamber. The sensitized plate, which has to be exposed to the light, and to respond to its most delicate operations, must be generated in darkness, in the dark chamber. This space, surrounded by bottles and boxes, and crammed with instruments, is the narrow world of the photographer, out of which he issues only for a few minutes into the light of his atelier, in order to return directly with his illuminated plate, and to subject this to various other chemical operations.

Many persons believe that the opening and shutting

of the slide — the cover or lid of the apparatus falsely called the machine—is the chief work of the photographer. Nay, it is related of a certain queen, that she thinks she is photographing, when she has all the necessary apparatus brought and prepared, and then, when all is ready for the result, opens and shuts the lid of the objective—a work that a child of five could do equally well. But this operation is only a link in a great chain of twenty-eight operations, through which each plate must pass to produce even a negative, while at least eight further operations are required to throw off a positive from this negative.

Let us look a little closer at these operations. The appearance of a dark chamber is by no means inviting. Even where the greatest order prevails, drops of solution of silver are diffused about, and black spots produced here and there. To this must be added a permanent odour of the vapour of collodion, and an unavoidable dampness from the necessary washing of the plates,—and all this is seen in the hazy *chiaro-oscuro* of a gas or petroleum lamp provided with a yellow shade, or of a small window fitted with a similar glass shade.

The remark must here be made at the outset, that the dark chamber of the photographer is not really completely dark. The light of *day* only must be excluded from certain operations; but the yellow light of the lamp is innocuous.

From this we learn the important distinction between light chemically operative, and light chemically inoperative. The light of the sun and of the blue heavens, the electric light, and the magnesium, are chemically very operative, gas light and petroleum light very slightly so;

whilst the yellow light of a spirit lamp, whose wick has been rubbed with kitchen salt, is entirely inoperative. The operative light of day, furthermore, can be rendered inoperative if allowed to pass through a yellow or, better still, a reddish-yellow glass shade. The light, therefore, that falls through the yellow window of a dark room is chemically inoperative, or in so slight a degree operative that it no longer causes any disturbing effect. It is remarkable that the yellow light which affects our eyes so powerfully, should influence the photographical plate hardly at all. Up to the present time this fact has not been sufficiently explained. It has disadvantages for practical photography; for example, a yellow garment becomes easily black in photography, a yellow complexion, yellow spots—such as freckles—appear almost black in the picture. Nevertheless, these disadvantages can be obviated by employing the negative *retouche* described at a future page. On the other hand, the inoperative property of yellow light has also its advantages for the photographer. It permits him to prepare the sensitive plates in a subdued light which does not injure them, and yet suffer his eyes to control the work. If the plates were sensitive to all kinds of light, it would be necessary to prepare the plates in absolute darkness, which would be very inconvenient.

The first operation required in preparing a sensitive plate—an operation which requires great care—is the cleaning of the glass. The slides, after being cut by the diamond, are placed some hours in a corrosive fluid —nitric acid—and by this means all impurities adhering to the surface are destroyed.

The acid adhering to it is removed by washing, and the plate is then dried with a clean cloth. To the uninitiated it would then appear perfectly clean, but the photographer subjects it to further polishing, by rubbing with a few drops of spirits of wine; or, still better, of ammonia. Each touch with the finger or rub of the sleeve of the cleaned surface, each drop of saliva which might chance to escape from the mouth in coughing, would spoil the polished surface; nay, even the atmospheric air produces with time disadvantageous effects. If a cleaned plate is left only twenty-four hours in the air, it gradually attracts its exhalations, and another cleansing is rendered necessary.

The cleaned glass is saturated with collodion. The collodion itself is, as we know, a solution of gun-cotton in a mixture of alcohol and ether; to which metallic iodine and bromine—for instance, iodide of potassium and bromide of cadmium—have been added. This solution must also be produced with the greatest attention to cleanliness; and, in order to preserve the purity of the materials employed, the mixture must be allowed to stand a long time, and the sediment carefully cleared of all fluidity. The coating of a plate with collodion is an affair of dexterity, and only succeeds with those who have witnessed the process and after some practice.

It is usual to hold the corner of the plate with the hand and to pour over the centre of it a circular mass of the thick fluid, and then to allow this to flow to all of the four corners by a gentle inclination of the plate in different directions, and thus ultimately to let the fluid flow off at one of the corners.

A considerable part of the fluid originally poured

upon it—that is, nearly half—remains, and adheres to the plate.

In the process of flowing off, streaks are usually formed, which would likewise spoil the picture; and therefore the plate, whilst being drained, must be constantly kept in motion until the last drop has run off. The fluidity stiffens into a soft, moist, spongy film. At the moment when this thick film has become stiff, the plate must at once be immersed in the solution of silver (silver bath).

And now a somewhat unusual action of the fluids takes place, for the film of collodion repels like fat the watery solution of silver, and a steady agitation of the plate in the solution is necessary in order to make the solution adhere to the plate.

This mechanical operation is accompanied simultaneously by a chemical change. The salts of iodine and of bromine that exist in the collodion film change their properties with nitrate of silver, and give birth to iodide and bromide of silver, and to nitric-acid salts. This iodide and bromide of silver colours the film yellow; and it is only now that the plate is prepared which serves as the ground of the picture about to be painted by the light.

All of these operations must precede the taking the photograph, and they are begun, in fact, at the moment when the person enters the atelier, and with proper management the plate is prepared before the arrangement for taking the portrait is concluded.

This arrangement is a labour of itself; and it is of a genuinely artistic nature. The points to which the photographer has to attend include a natural and yet graceful attitude of the original; the choice of the side which presents the most advantageous aspect; the picturesque arrangement of the dress; the removal of inappropriate objects which ought not to appear in the picture; the addition of those that are suitable, such as a table, a cabinet, or a background; lastly, an appropriate direction of the light. Only a few minutes can be devoted to these arrangements, for persons cannot endure long delays or experiments; and the plate itself only lasts a short time in the sensitive condition, for it is damp through the adhering solution of silver, which soon dries up, and the plate is then useless.

When the exposure to the light has been accomplished—a process during which the person being photographed must remain perfectly quiet—the sensitive plate is brought back into the dark chamber.

For the purpose of transporting the plate, which must evidently be guarded very carefully from the daylight, the photographer employs a little flat box (Fig. 10), called the cassette, whose floor, H, and cover, D, can be drawn out and closed. In the corner there are silver ledges on which the plate lies; a wedge fastened to the upper lid keeps them in their place. Thus they can be easily carried in the closed cassette and placed within the camera obscura; there its ground-glass slide is moved to and fro until the picture clearly appears upon it. After the exposure to light has taken place the plate is taken back in the cassette to the dark chamber.

And now follows one of the most important operations, the development of the picture. Upon the plate there is as yet no trace of a picture visible. The operation of the light consists in quite a peculiar change of the iodide of silver which forms the principal constituent of the plate. This iodide obtains through the light the property of attracting pulverized silver, if this has been precipitated on the plate in any shape. This precipitate is produced by the following operation. If a solution of silver is mixed with a very diluted solution of green vitriol, there results by slow degrees a precipitate of metallic silver—not, however, as a green shining mass, but as a grey powder. A solution of silver adheres

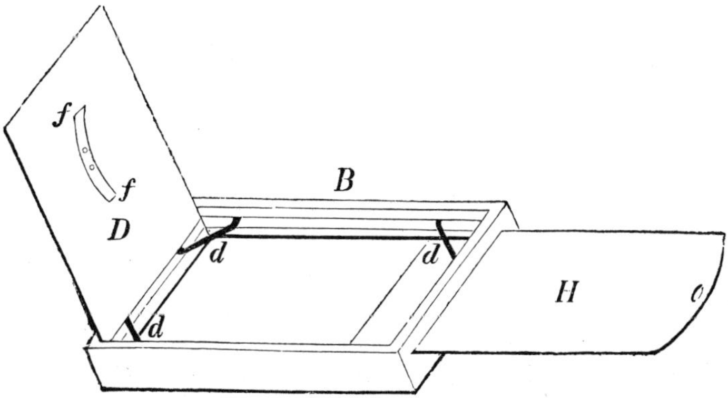

Fig. 10.

now to the sensitive plate resulting from the bath. If after this a solution of green vitriol is poured upon it, a silver precipitate is also occasioned, and the picture is seen suddenly to make its appearance, by the silver powder adhering to the part exposed to light.

The features of a portrait that are first visible are the lightest—the shirt, then the face, and lastly the black coat. The negative thus obtained, however, is by no means completed by the operation.

The picture is usually too attenuated to answer in the

positive process for the production of a paper impression with the help of light; for the production of such an impression results from the light shining through the transparent places of the negative, and colouring dark the places underlying them, while it is repelled from the parts which have to remain black. The parts in question of the negative must be sufficiently transparent to produce this effect.

The impression must, therefore, be more strongly defined; and this takes place by repeating the developing process. A mixture of green vitriol and of a solution of silver is poured upon the picture, and a silver precipitate is formed again on it, adhering only to the lines of the picture, and therefore giving these an intenser colouring. If the plate is not perfectly clean in the processes of developing and defining, silver is precipitated upon the dirt stains and produces spots. After the defining of the picture, or the so-called fortifying process, has been completed, it is only necessary to remove the iodide of silver, which diminishes the transparency of the clear parts of the plate. Then a solution of hypo-sulphite of soda is poured on the plate. This salt has the property of dissolving insoluble salts of silver, so that the iodide of silver vanishes under the influence of this solution. This is the fixing process. Lastly, the plate is washed and dried. If it be borne in mind that all these different operations are performed on a little film, liable to be injured by the least contact of any kind, it is not to be wondered at that in treating matters of such a delicate nature the inexperienced beginner has to destroy so many coatings before a proper one is prepared.

Even when dried, the picture is very liable to injury;

and therefore photographers, in order to protect it, cover it with a varnish, that is, with a solution of a resinous nature, such as shell-lac and orpiment, or red arsenic in spirits of wine. The fragile glass negative is thus brought to completion.

This sketch of the operations which a photographer is obliged to carry on in order to produce a negative, is sufficient to show that photography is a more difficult art than some persons imagine, and that it requires something more than the opening and shutting of a lid.

The chief requisite for the success of these operations is routine, that is, the unfailing accuracy obtained in practising each part of the process. Faults that are made in any particular operation of the process are, as a general rule, irremediable; and therefore it is absolutely essential to avoid them.

CHAPTER VI.

THE POSITIVE PROCESS.

Character of the Negative—Departure from Nature—Negative *Retouche* —Production of Sensitive Paper—Striking off Impressions—Toning with Chloride of Gold—Fixing—Cause of Fading—Quantity of Silver in the Picture—Toning down of Photography.

In the preceding chapter we have become acquainted with the production of a negative from nature. However interesting such a negative might be, nevertheless, it could not satisfy the purchaser of a portrait, because it showed everything reversed. The white face it made black, and the black coat, light. No one would hang up on his wall a picture representing him as a Moor. It was therefore necessary to obtain a positive impression from this negative. We have already learnt how this is effected in the chapter on the "Licht-paus" process. It is the old Talbot method that is here employed. But we must still mention some very important collateral operations which are of high significance in modern photography.

The camera, the negative process, and the photographer who knows how to manipulate intelligently, no doubt produce a negative which, laid over sensitive paper and exposed to the light, yields a positive; but although this positive is very faithful in the delineation of figures—that is, of their outline—it yet presents marked departures from nature. It is especially evident that the relations of light and shade are by no means correctly given. In general the light parts appear too light, the dark parts too dark—as, for example, the folds in a dress, the skin, and, moreover, the shading under the eyes and chin. When photographers knew nothing of art, these defects were taken as a matter of course. People protested that photography was correct because nature, through photography, was herself the artist. But in this conclusion the co-operation of the photographer was overlooked.

No doubt nature that is, the object to be taken—makes an impression upon the plate, by the light issuing from it; but an impression of light is not a picture—it is indeed, of itself, invisible; nay, more, the strength of the impression of light is entirely at the discretion of the photographer, who can make it weak or intense by a greater or less exposure. There is no rule which determines the length of time a photograph has to be exposed to the light.

The fact is that nature, properly speaking, only determines the outline of the picture, while the relations of light and shade depend partly on those distinctions of nature, and partly on the good pleasure of the photographer.

The print or impression of the light is developed; hence it becomes visible, and finally the developed picture is brought more strongly out. By this means the photographer can at his option increase, and even exaggerate, the contrasts of light and shade.

If the negative is carefully compared with the original, we shall find that many dark parts have not appeared at all, because the exposure was too limited for them to produce an impression upon the plate; others have appeared, but too indistinct. On the other hand, very clear parts—for instance, the shirt-collar—have an excess of clearness and whiteness, and the needlework upon it is invisible because the time of exposure was too short. In the case of long exposures it is often remarked that clear parts differing little in colour are entirely confounded, that is to say, form a single white patch.

Moreover, the accessories which a painter would undoubtedly omit, such as warts, pockmarks, little hairs, are all as clearly defined as the principal features; and thus the negative is neither a correct nor an agreeable repetition of the reality, but produces in the positive a picture which shows considerable departures from nature, and is often inaccurate by giving too much prominence to accessories.

In the first period of photography these departures were overlooked. Every one was content to possess a portrait which at least showed the outlines correctly; and what was defective in the negative it was sought to atone for through the retouch of the positive. But this retouch rendered the picture dear; and as it began to be the custom to order pictures by the dozen, the endeavour was made to evade this labour, which had to be applied to each individual picture, by carrying it out in the negative.

A single touched-up negative gave hundreds of unexceptionable impressions which did not require to be retouched, and thus the negative *retouche* became the first and most important operation to produce a faithful and agreeable picture. The essential characteristic of this negative *retouche* consists in entirely covering several parts. For example, the freckles and warts in the clear negative are entirely removed by the pencil, or Indian ink. Other parts—for example, the too delicate details of the hair—are more defined by pencil strokes. Many shadows—for example, the wrinkles in the face—are softened off by slight touches of Indian ink. This labour must always be carried out with the thought that all which the painter draws on the negative with his black-lead pencil will appear the opposite; that is, light in the positive.

It is requisite, therefore, for the negative *retouche* to have a thorough knowledge of working with lead-pencil and Indian ink, to render the different shades in the positive process. The best draughtsman and painter is on that account still far from being able to retouch a negative.

It is to be remarked that the negative *retouche* may, under certain circumstances, go too far. By covering each wrinkle he can make an old face young; he can beautify an ugly original by cutting away a hump on the back, or other abnormal growths; and these tricks are often put into requisition for the vanity of sitters, and are dearly paid for.

Plate VI. (p. 245) represents two portraits of the same person, one after a retouched negative, the other after a negative that had not been retouched. They represent a celebrated singer (Mdlle. Artot); the spots on the skin and the dark shadows, on the picture which is not retouched, are clearly to be seen, while in the retouched one they are not visible.

In many cases the negative *retouche* is a concession to human vanity, but this is by no means always the case.

As already explained above, photography does not always reflect correctly the natural colours. Yellow often becomes black, and blue, white. Therefore, in producing a picture of brighter hues, photography is often very deficient in the reproduction of their tones. Then the negative *retouche* is a powerful aid to correct this fault, and through this alone have photographs taken from oil paintings attained their present perfection. We will treat of this subject in a future chapter. Let us now consider the operations of the positive processes.

The first operation is the production of the sensitized paper. A piece of paper coated with white of egg and moistened with a solution of kitchen salt is laid in a cup with a solution of silver. The paper floats upon the liquid, it sucks it up, and chloride of silver is formed, through decomposition with the kitchen salt. At the end of a minute the paper is taken out of the silver solution.

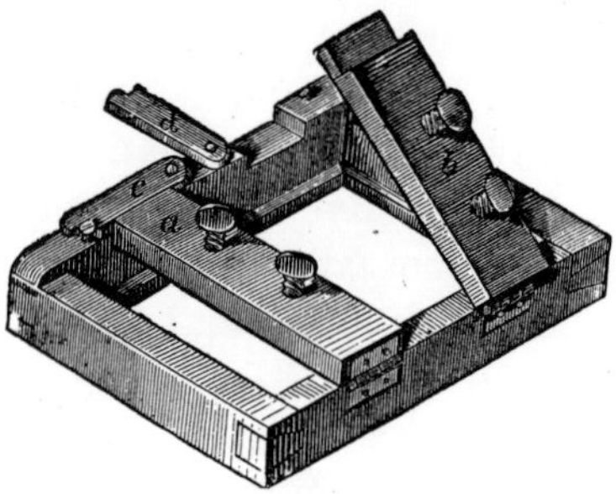

Fig. 11.

The wet paper is but slightly sensitized; it becomes fully sensitized only after being dried. The dry paper, saturated with chloride and nitrate of silver, is then pressed together in the copper frame (Fig. 11), which is similar to the one described at a previous page. Then the whole is exposed to the light. The same process ensues which we have described in the chapter on "*Licht-paus* paper"; the light shines through the clear places of the negative and colours dark the paper lying under them, but the paper under the dark places of the negative remains white, while it assumes a slight colour under the half-tones. In this manner a faithful positive reprint of the negative is produced, presenting a beautiful violet-brown tint. We know from the description of the *licht-paus* process, that this reprint would not stand the light long, because the paper is still sensitive to light. The salts of silver contained in it must be removed if the impression is to be made lasting. To this end a solution of hypo-sulphite of soda must be employed. If the impressions be plunged in this solution, they become durable in the light; but, unfortunately, by thus dipping them, they suffer a peculiar change of colour, assuming an ugly brown tint. This tint is no injury in technical and scientific pictures, but detracts greatly from portraits and landscapes; and in order to give these a more agreeable tint before fixing them, they are plunged into a diluted solution of chloride of gold. This process is called toning down.

In this operation a part of the gold is precipitated on the outlines, giving to these a bluish shade; and now, after plunging it into a solution of fixing sodium, the tone of the picture is not essentially altered.

The picture thus produced consists partly of gold, partly of silver, in a finely powdered state, and only requires to be thoroughly washed in order to establish its durability. If this washing is omitted, small particles of fixing sodium holding sulphur in suspension remain behind, and these become decomposed and form on the picture yellow sulphide of silver. This accounts for the fact that the pictures of an earlier period, when from ignorance of these results this thorough washing was neglected, so often turned out pale and yellow.

It is surprising what a small amount of silver and gold is required to give an intense colour to a whole sheet of paper. For in a sheet of this description—1·447 feet by 1·546 feet—which has become completely blackened, there are only 1·3162 grains; whilst in a picture of this size there are only 0·075, that is, about one-thirteenth of 15·440 grains,* and in a carte de visite one-five-hundredth of 15·440 grains† of silver.

It must be here remarked that pictures which are fresh when printed, pale a little in the fixing process; and hence the photographer usually prints these darker than they ought to remain. Accordingly, the printing process requires a practised eye, simple as it may appear.

In certain cases tricks of art are employed to produce agreeable effects, and among these is that of toning down. Our readers are no doubt well acquainted with portraits on a white ground, the outlines of which gradually become confounded with the ground tint of the picture. This effect is produced in a very simple manner by placing what is called a mask on the copying frame. This mask is a piece of metal or cardboard (Fig. 12) in which an oval hole *b* is cut. This is placed on the copper frame *K K*, so that the part of the negative which shall be impressed on the picture lies perpendicularly under it. This part is then affected

* One-thirteenth of a gramme = 1·187 grains.

† One-five-hundredth of a gramme = ·031 grains.

perpendicularly by the broad bundles of light *S S*, and intensely coloured, while the collateral parts lying under the mask are affected only by the small patch of light, *S′ S′*. Accordingly, they only give a pale reprint on the paper, as they are remote from the margin of the mask. Thus a gently vanishing margin is produced, looking very artistic, and yet only the result of a very simple trick of art.

The picture which the photographer produces in the manner now described only requires some rectifying to be an elegant drawing-room ornament. It is cut in a regular shape, square or oval, and fastened with clean paste to white cardboard, and finally, after drying, and the removal of little blemishes, by slightly touching up

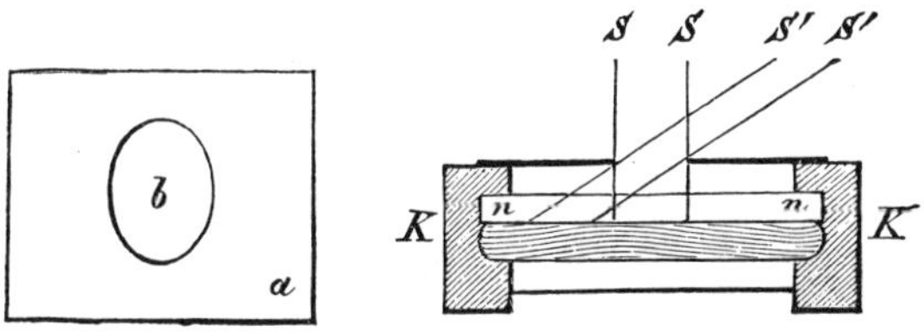

Fig. 12.

with the paint brush, it is rendered glossy by two smooth steel rollers, and receives a satin-like surface.

Certain sizes have been adopted through custom by the public. Among these is the shape of what is called the "carte de visite," and "cabinet size." The former is rather larger than an ordinary visiting card. The latter is two and a-half times as large.

The carte de visite was introduced at Paris by Disderi, in 1858, speedily secured admirers, and has been diffused over the whole earth. Even chemical photographers prepare photographs in the form of cartes de visite.

The carte de visite and the cabinet form—first adopted in England, and a great favourite in America—are not confined to portraits, but also employed for landscapes and photographs taken from oil paintings. Millions of these pictures are sold every year, and a properly arranged album for preserving them is found in almost every family.

Photography admits of such small forms because of its fine details, but it is by no means confined to them. It freely admits surfaces that take in a portrait of life-size. The production of the latter necessitates a peculiar process, called the enlarging process, which will be treated of in a future page.

CHAPTER VII.

LIGHT AS A CHEMICALLY-OPERATIVE AGENT.

Theory of Photography—Nature of Light—Undulatory Theory—Semi-Tones—Mouldering away of Red Sulphuret of Arsenic—Chemical Decomposition by Light—Colours and Tones—Their Vibrations—Refraction—Dispersion of Colours—The Spectrum—Spectral-Lines—Invisible Rays — Photographs of Lunar-Landscapes — Abnormal Photographic Effect of Colours—Photography of the Invisible.

GOETHE says, "All theory is grey, and the golden-tree of life green." This saying has often been misunderstood and abused, especially by those too lazy to think; but, faithful to its true meaning, we have first treated of a multitude of facts from life—that is, from the history and practice of photography,—and now we proceed to describe, by the help of science, how and why, not the golden but the silver tree of photography becomes verdant, blooms and bears such splendid fruit.

Two sciences join hand to accomplish the wonders of photography. One is Optics, a division of Physics, and the other Chemistry. We have already shown that they alone are inadequate to fulfil the requirements for the production of a photograph. Æsthetical claims have to be considered; and thus photography unites in itself the provinces of natural science and of the fine arts which seem remote and incapable of union. We shall attend first to the optical principles—that is, to light—as the force which occasions the chemical changes in photography. We shall see that its chemical operations have not only become the basis of our art, but that they have played, and still play, a still more important part in the development of our planet.

We are aware of the existence of sun, moon, and planets. We know their distance; nay more, we know their elements, though we are separated from them by millions of miles.

We are indebted for all this knowledge to light. What is light? An undulation of the ether. And what is the ether? An infinitely delicate fluid, which fills all the space of the universe, and undulates like all fluid. If we throw a stone into water, waves are produced—that is, circles or rings, or hills and valleys, are formed; these appear to widen out from a centre, and as they extend become gradually less, until they finally disappear. If several little stones are thrown at the same time into the water, each of them forms its own system of undulations. These intersect each other in the most complicated manner; and, although a confusion of rings takes place, it is wonderful that none of them disturbs the other, and that each circle widens out regularly from its own centre, where the stone fell into the water. (See Fig. 13.)

If a handful of sand, which contains many thousand grains, is thrown into water, and if the attention be directed to the undulations of a single grain, it will be probably remarked that this one, without being affected by the countless other waves, widens out into a regular circle.

These undulations are one of the most remarkable movements in nature, taking place not only in water, but in the air, where they occasion the propagation of sound.

The peculiar feature of the undulatory movement consists in the fluid appearing to advance without really doing so. If, sitting on the side of a sheet of water, we see an undulation approach, it appears exactly as if the particles of water were approaching us from the origin of the movement.

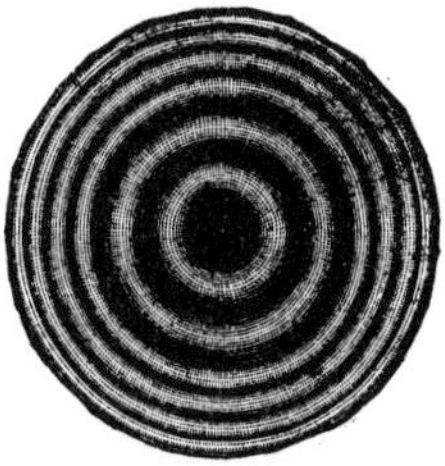

Fig. 13.

It is easy to prove that this is an error by throwing sawdust or a piece of wood into the water. It dances up and down upon the ripples without moving from the spot. Indeed, the undulation is itself only an up and down motion of the particles of the water, and this movement it communicates further and further to the neighbouring particles of water.

Exactly in the same manner light spreads in undulations from a luminous body through the ether of space in all directions. The movement of the undulation we call a ray of light. We perceive it as soon as it reaches our eye, whilst the vibrating ether strikes our retina.

Now, we know that the undulations of tone are able to set other bodies in motion. If the A or second string of a violin is struck, the A string of a piano standing near sounds distinctly with it. Nay, even if the damper of a piano is raised and any tone made to vibrate, instantly the string of the violin sounds which has a similar tone. The same thing happens with a glass bell of the same tone. There are people even who can make a glass break by a shrill tone of their voice. The glass is so shaken by the violent undulations of the air that it falls to pieces. Under such circumstances, it need not surprise us that the undulations of light agitate bodies so forcibly that they fall to pieces.

Red sulphuret of arsenic offers the most remarkable example of this kind. This is a beautiful mineral of a ruby red colour, in the form of splendid crystals, which consist of sulphur and arsenic. If a crystal of this kind be exposed for months to the light, it becomes pliant and falls into powder; and in this way many very fine pieces of this beautiful mineral have been lost in the mineralogical museum of Berlin.

This is only a mechanical, and not a chemical, operation of light; but it gives an insight into its chemical working. Heat occasions chemical decomposition by extending bodies, and thereby removing their atoms so far apart that the chemical power which unites them loses effect, and the component parts separate. Thus the oxide of mercury is by heat resolved into its parts, mercury and oxygen.

This decomposition is effected by light when the atoms of a body are agitated by its undulations, that is to say, are made to vibrate; and if these vibrations are unequal, a separation of the parts takes place, and the body falls to pieces.

The undulations of light are not a fiction. Not only has their existence been ascertained, but their size has been determined. The latter is extremely minute, but nevertheless is susceptible of measurement.

The waves of sound and the waves of light have therefore a certain analogy; and as there are different tones in music, so are there different tones in light. The number of tones is great. The simplest piano has nine octaves, and there are tones below and above it. But the number of colours is small; only seven of them can be distinguished—red, orange, yellow, green, blue, dark blue, and violet,—the well known colours of the rainbow. The painter, indeed, contents himself with three ground tints—yellow, blue, and red. All the others are the result of their mixture; and the larger scale of colour of the painter consists not of simple tones of colour, but of what may be called chords of colour.

The deep tones of music give few undulations, the higher tones more. For example, an A string makes 420 vibrations in a second, the A an octave lower makes 210, the great A 105.

In light, red is the colour which gives the fewest vibrations; it is the lowest tone in colours, and violet is the highest, giving vibrations twice as rapid as red. With regard to tones, we know that they all spread with equal rapidity in the air; if this were not the case, a piece of music would sound in the distance as the most disagreeable discord.

It is the same in the kingdom of light—the colours, without exception, are propagated through the ether with equal rapidity, the red as fast as the violet. But, whilst the reverberation in the second passes over only 1024 feet = 333 meters, in the same time the light hastens 42,000 miles, and the deepest tone of colour—red—traverses in a second 420 billion of vibrations; that is to say, a million times million as many as the tone which is marked in music with a bar over the ā,* that is,

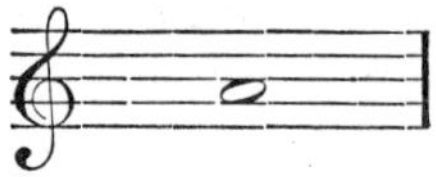

The small number of the colour-tones compared with the large number of musical tones is very striking. But the fact is, that, besides the seven invisible colours, there exist invisible shades, which lie partly above and partly below the visible colours.

These invisible colour-tones are partly disclosed by the thermometer, which reveals the lower tones, and partly by substances sensitive to light. For it is remarkable that the colour-tones, which are higher than the violet, though invisible, have a powerful chemical effect.

We name the invisible tones of colour above violet, ultra-violet, and those beyond red, ultra-red.

In the common white light all the tones of colour are found together, and in combination they excite the feeling of whiteness; but if we wish to consider the tones of colour separately, we must part them, and this is done by the help of a prism.

Every polished crown-glass prism causes the rays seen through it to appear like a rainbow streak containing the primitive colours we have named above. This separation of the colours in the prism takes place by refraction.

If a ray of light passes from one transparent body to another, it is deflected from its rectilinear direction, and this deflection is named refraction.

* We may here remark that the tone ā is not everywhere the same. The ā of the Berlin Opera is the highest; it has 437 vibrations,—the Italian Opera at Paris only, on the contrary, 424 vibrations. We have adopted for the sake of simplicity a round number, 420.

For example, if the ray *a n* (Fig. 14) strikes a watery surface, it does not continue in its original direction *a n*, but in the direction *n b*. If at the point *n*, where the ray falls into the water, a perpendicular line *n d* be raised, this is the plumb line; and the rule is, that if a ray passes from a thinner medium (for example, air) into a thicker one, it approaches the plumb line, for *n b* is evidently nearer to the plumb line than *n a*. It is otherwise if a ray passes from a denser to a thinner medium,—for instance, from glass into air,—then the ray *n b* departs from the plumb line *n d*; that is, the angle which it makes with the plumb line after refraction is greater than the angle which it makes with it before.

Now, it is a remarkable fact that the light of unequal shades of tone is refracted also unequally.

If a bundle of white sun's rays is suffered to fall on a piece of glass, the violet rays are deflected in a greater degree than the blue rays, the blue more than the green, yellow, and red; and the result of this is that the white bundle of rays is decomposed into a rainbow-coloured fan, violet, indigo, blue, green, yellow, orange, and red.

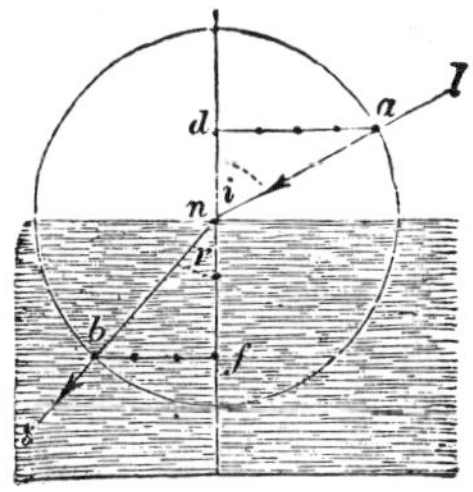

Fig. 14.

This phenomenon is the cause of the rainbow. If a ray *a* falls on a drop of water (Fig. 15), it is refracted and at the same time divided into a coloured fan, which is reflected from the lower part of the drop, suffers again a refraction and dispersion of colour *b*, and issues as a broad bundle of colour. In open daylight this cannot be clearly seen, because our eyes are dazzled by the clear light surrounding them. In order to observe the pure colours of the spectrum, it is best to place it in a darkened room, in which the light is allowed to enter only through a small slit (*b* Fig. 16).

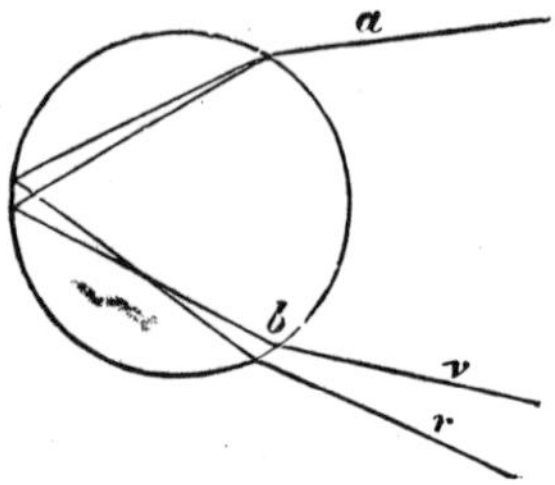

Fig. 15.

The prism *s* is placed on a line in front of the chink, when the colours of the spectrum are clearly seen upon the opposite wall. If the chink is sufficiently narrow, a row of dark lines is observed within it, which cut through the coloured stripes perpendicularly.

These lines were first seen by Wollaston, studied more exactly by the celebrated Frauenhofer, and called after him Frauenhofer's lines.

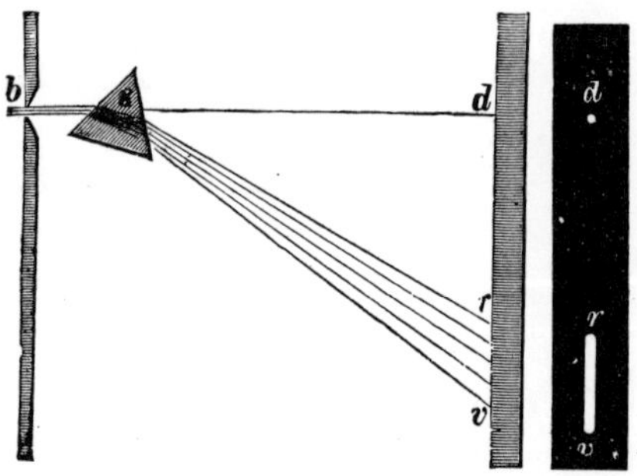

Fig. 16.

The lines are always found on the same spot, so that they can be considered as natural music lines, upon which the scale of colour is written; and as the music lines serve for the recognition of the musical tones, so do the lines of the spectrum indicate the exact places of the scale of colour.

If we use the term green in the spectrum, this would be a very vague designation; whereas by presenting the line on the spectrum in which green is found, its place is at once made known. To this end certain characteristic names were given to the lines by Frauenhofer, which he indicated by letters; a certain line in the red he called *A*, another in the yellow *D*, one in the violet *H*, and *H'*. As the number of lines reaches several thousand, these letters do not suffice to indicate them all. (See Fig. 17.)

The lines thus named are found especially in the sunlight; the light of other stars commonly shows other lines. The light from artificial sources does not show dark, but bright lines; a flame coloured yellow with kitchen salt shows, for example, a very characteristic line in the yellow; a burning magnesium wire shows more blue and green lines.

The situation of these lines agrees exactly with that of certain dark lines in the spectrum. For example, the yellow line in a flame coloured with kitchen salt exactly coincides with line *D* in the spectrum. The green lines in a flame of magnesium coincide exactly with lines *E C* in the spectrum.

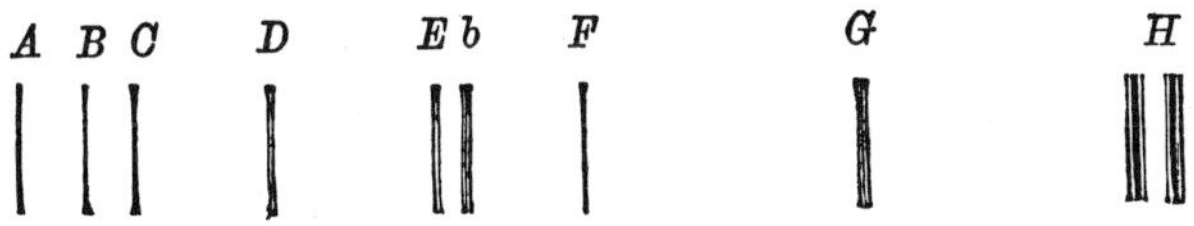

Fig. 17.

This remarkable coincidence led to the surmise that the lines in the sun's solar spectrum might owe their existence to the same substances that produced the coinciding lines in earthly flames. Kirchhoff converted this surmise into a certainty, and was thus able to determine from the lines in the solar spectrum the substances present in the red-hot body of the sun, and thus to demonstrate the chemical composition of a star distant more than 90 millions of miles (spectrum analysis).

But the spectrum contains still other numbers, which have not been discerned by the human eye, but by the photographic plate.

If a sensitized plate be exposed to the operation of the spectrum, it is observed that red and yellow make only a very feeble impression upon it. Light blue produces more effect, but dark indigo and violet the most; and in the space where no rays can be perceived by our eyes, a distinct impression is produced, and extends beyond violet for a space almost as long as the visible part of the spectrum.

From this fact the existence of the ultra-violet rays was ascertained. Accordingly, the retina of our eye and the photographic plate show an entirely different susceptibility. Our eye is affected most powerfully by yellow and green. These colours appear to us the clearest, while the photographic plate is not at all affected by them; but it receives powerful impressions from indigo and violet rays, which appear dark to our eye, and even from rays which to our eyes are invisible.

Therefore it is natural that photography should represent many objects in a false light. Further back we called attention to the fact that photography is much less sensitive to feebly lighted objects (objects in shadow). This is most clearly seen in the fact that the eye can easily perceive objects in the moonlight, which is 200,000 times weaker than that of the sun; whereas the photographic plate of a lunar landscape is not able to produce any picture at all. The photographic lunar landscapes sometimes offered for sale have been taken in the daylight and copied very darkly, so that they produce the effect of moonlight. These pictures are very popular at Venice.

This small susceptibility of the photographic plate to feeble light explains the reason why shadows in photography are generally too dark. To these defects must be added the false working of light,—blue generally works clear, yellow and red work like black. The yellow freckles appear therefore in a picture as black spots, and a blue coat becomes perfectly white. Dark blue flowers on a light yellow ground produce, in photography, light flowers on a dark ground. Red and also fair golden hair become black. Even a very slight yellow shade has an unfavourable effect. A photograph from a drawing is often blemished by little ironmould specks in the paper invisible to the eye. These specks frequently appear as black points. There are faces with little yellow specks that do not strike the eye, but which come out very dark in photography. A few years ago a lady was photographed at Berlin, whose face had never presented specks in photography. To the surprise of the photographer, on taking her portrait specks appeared that were invisible in the original. A day later the lady sickened of the small-pox, and the specks at first invisible to the eye, became then quite apparent. Photography in this case had detected before the human eye the pock-marks very feebly tinged yellow.

In the photographs of paintings, such abnormal workings of colour are still more evident, and can only be removed by appropriate negative retouches.

It is proper to observe, however, that by no means all shades of blue become light in photography. For example, indigo forms an exception, appearing as dark as in nature, and this is shown in the photographs of the uniforms of Prussian soldiers. The reason of this is, that indigo contains a considerable amount of red. On the other hand, cobalt blue and ultramarine produce almost the effect of white. Again, cinnabar red works dark, also English red; whereas madder red, which contains blue, becomes very light. Chrome yellow becomes much lighter than Naples yellow, Schweinfurt's green becomes lighter than cinnabar green. No one of our pigments contains a perfectly pure spectrum colour, but consists always of a mixture of different colours, and therefore is essentially modified by photographic operations.

If the effect of the colours of the spectrum on photographic plates is more narrowly examined, it is observed that indigo produces the greatest impression. Nevertheless, the differently sensitized photographic preparations offer somewhat different results in this respect. Chloride of silver is most sensitive to violet, but non-sensitive to blue. Bromide of silver is also sensitive to green, and iodide of silver only to violet and indigo. Mixtures of iodide and bromide of silver are only sensitive to blue and green. The writer of this work succeeded, in the end of 1873, in making photographic plates sensitive even to those colours that were before considered to be inoperative, *i.e.* yellow, orange, and red. He found that if certain coloured substances that absorb light were added to bromide of silver, which is by itself too little sensitive to green, the sensitiveness of this bromide to green is considerably increased. In like manner, if coloured substances absorbing yellow or red light are added to it, they make bromide of silver sensitive to yellow and red light. After this discovery, we may hope that the difficulties attending the taking of coloured objects may be soon overcome.

Mention has often been made of the photography of the invisible. The cases already recorded of the photographs of invisible pock-marks belong to this. But the photography of an invisible quinine writing is especially understood by tho term photography of the invisible. If a writing is made on paper with a concentrated solution of sulphate of quinine, the result is scarcely visible. If this is photographed, it appears black and plainly visible in the picture. The sulphate of quinine has the property of lowering the tone of violet, of ultra-violet and blue rays; that is, of converting them into rays of less refraction and of less chemical effect; therefore the light issuing from quinine produces little or no effect, and the written characters become black.

This property of the sulphate of quinine serves also to make ultra-violet rays visible. If a piece of paper that has been rubbed with sulphate of quinine is held in the spectrum, the originally invisible ultra-violet part of tho spectrum is seen to shine in the bluish green light.

Other substances produce this effect, such as uranite, Devonshire spar (fluor), and therefore this property has received the name of fluorescence.

CHAPTER VIII.

CHEMICAL EFFECT OF DIFFERENT SOURCES OF LIGHT.

Artificial Light—Magnesium Light—Lime Light—Electric Light—Representation of Subterranean Places by Reflected Sunlight—Chemical Intensity of Sunlight and of the Blue Sky Light—Breathing of Plants under the Influence of Light—Effect of Light in the History of the Development of the Earth and in the Economy of Nature.

From the facts explained in the foregoing chapter, it follows that chemical effects are chiefly produced by the ultra-violet, violet, and blue rays. It is therefore evident that a light, from whatever source, will produce chemical effects with an intensity proportioned to the amount of these rays it contains.

Lamplight, gas and petroleum light are very poor in such rays. Therefore these operate only feebly on the photographic plate, and photographers can prepare their sensitive plates in a subdued lamplight.

This is also done frequently in the day by allowing the light to pass through yellow glass.

The white Bengal light of arsenic, the flames of the blue Bengal light, and those of burning sulphur, produce a much more powerful chemical effect. The latter possesses only a small illuminating power, because it contains yellow and red rays, emitting little light; but, on the other hand, it is rich in blue and violet. Photographs have been actually taken by help of these flames.

But the above are greatly surpassed by the effect of the lime, the magnesium, and electric lights. The magnesium light is very simply produced by the burning of magnesium wire.

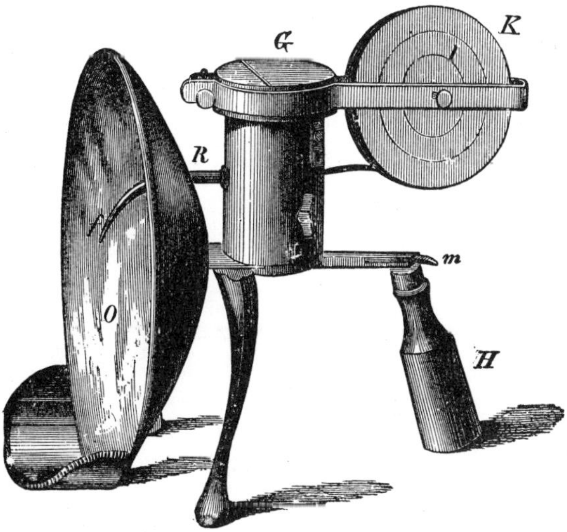

Fig. 18.

Magnesium is a metal which forms the chief component part of magnesia. Magnesia is nothing but magnesium rust; that is, a combination of magnesium with oxygen.

If magnesium wire is burned, it combines at a red heat with the oxygen of the air, precipitating the oxide of magnesium. The magnesium light is very convenient in its application. An ounce of magnesium wire, sufficient for fifteen to thirty experiments, can be easily carried in the pocket. But the general use of the light is impeded by the price of the metal (five groschen the gramme *) and by the smoke which it emits. The writer of this book has repeatedly employed it with success in taking the sculptures in the sepulchral monuments of Egypt. When burning the magnesium wire, Solomon's lamp is used (Fig. 18). This consists of a round vessel *K*, upon which the wire is coiled, a watch-work *G*, which conducts the wires by means of cylinders through the pipe *R*, at the top *f* of which the wire is lighted. The apparatus, by the concave mirror *O*, throws back the light as a parallel bundle of rays.

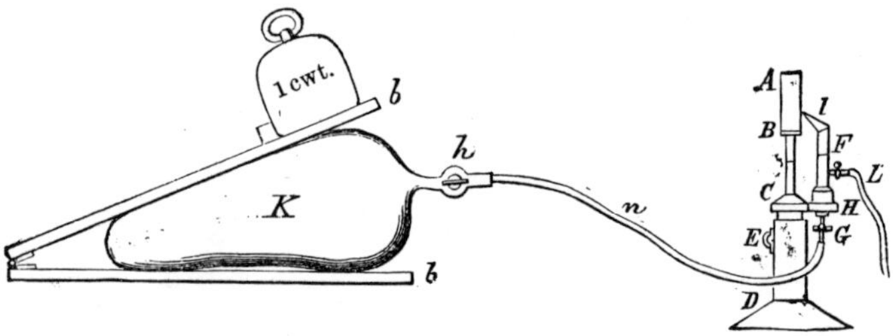

Fig. 19.

By means of the handle *H*, the lamp with its bundle of rays can be turned in any direction, and the watch-work can be instantly stopped by the key.

The magnesium light is surpassed in strength by Drummond's lime light. This is produced by a gas or spirit flame, into which oxygen gas is blown. The oxygen gas is produced by a salt rich in that element,

* 1 groschen = six-fifths of a penny, or $1\frac{1}{4}d.$ nearly. 1 gramme = the 1000th part of a cubic metre, about nine solid feet of water at the ordinary average temperature.

the chlorate of potassium. This salt contains the oxygen combined with a solid. Being heated, it escapes as a gas, and is received into an india-rubber bag. (See Fig. 19.)

This bag is closed by means of a stopcock, and when used is placed between two pieces of wood *b b*, a weight being placed on the upper one. By the pressure of this weight the oxygen gas pours through the cock *h*, and the india-rubber pipe *n*, into the oxygen lamp *D*. To this is attached a burner *H F*, running into the point *I*. The lighting gas which serves for combustion enters through the cock *L*, which is connected with a gas tube.

The combustion takes place at the point *I*. Without oxygen the lighting gas burns with a clear but soot-producing flame; but as soon as the oxygen is turned on, the flame becomes smaller and blue in colour, and burns with an intense heat.

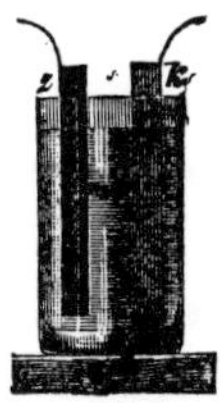

Fig. 20.

Its illuminating power is small, but as soon as the flame has brought the lime cylinder to a red heat, a dazzling white light issues from it, which has a very intense effect in photography, and has been used with success by Monckhoven and Harnecker to produce pictures on an enlarged scale.

The same apparatus serves for the production of what are called cloud pictures.

The electric light, produced by help of an electric battery, has a still more powerful effect than the lime light.

If a piece of cannel coal (*k* Fig. 20) and a piece of zinc are dipped together into an acid (diluted nitric acid or sulphuric acid), electricity is developed, which produces a spark on bringing together the two ends of the zinc and coal rising above the fluid; this spark is, however, very feeble. But if several vessels containing zinc cylinders *z* and pieces of charcoal *k* are employed, the spark becomes very intense; and, as we are able to increase to any extent the number of these elements, we are able to produce a cone of light of any degree of brilliancy, exceeding all other artificial light.

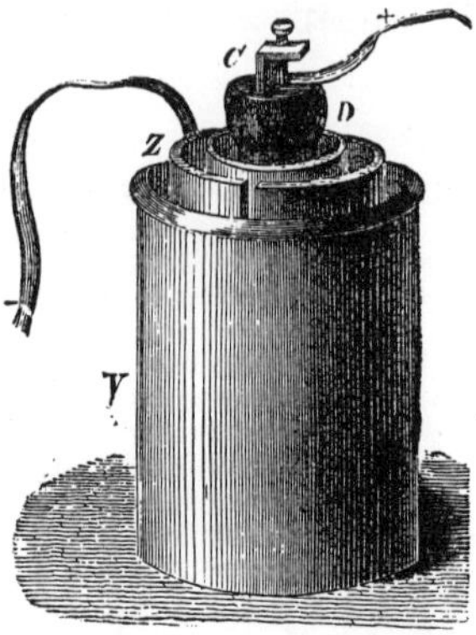

Fig. 21.

In arranging electric batteries of this kind, the zinc of one element is connected with the charcoal of the following, and the zinc of the latter with the charcoal of the third element. (See Fig. 22.)

If the two wires issuing from *Z* and *C* are brought together, a spark of light is produced by the electric stream, burning the metal wire.

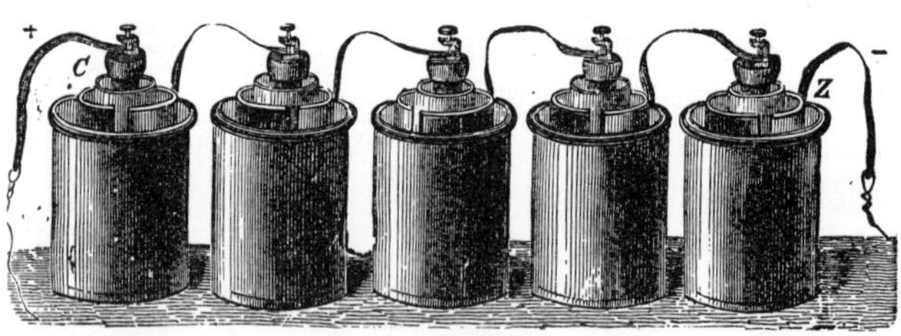

Fig. 22.

The light is generally produced between cones of charcoal placed in front of a concave mirror (Fig. 23).

The apparatus *S* and *S′* serve to approach or withdraw the cones, while the upper one is connected with the wire *K* by the foot *F*; the lower one is connected with the wire *Z* of the electric battery. Thirty-six elements

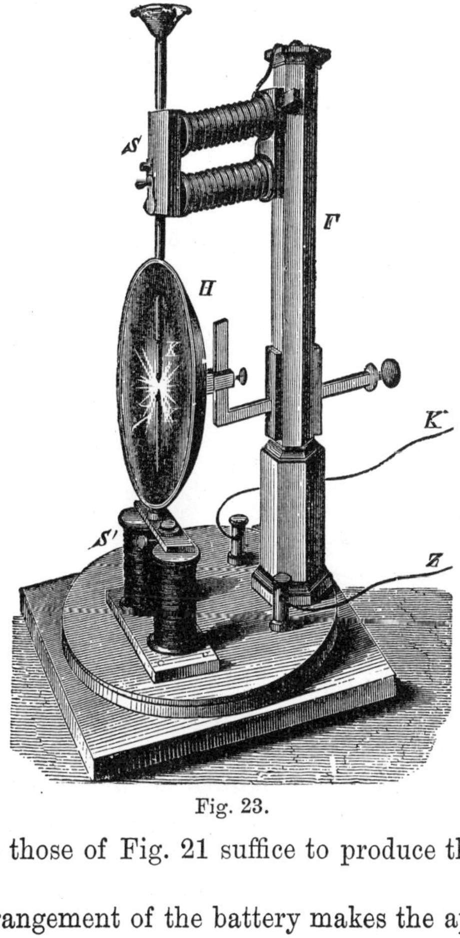

Fig. 23.

similar to those of Fig. 21 suffice to produce the electric light.

The arrangement of the battery makes the application

of the light inconvenient. In other respects this light surpasses all others in photographic effect.

Nadar has made with it many excellent pictures in the catacombs of Paris. It has also been used to take portraits. But in the latter case, the employment of such a dazzling artificial light is attended with the drawback of occasioning harshly defined shadows, that disfigure the portrait.

It has been attempted to evade this by allowing an electric light of less power to operate on the shaded side; but it is difficult under this dazzling light, as in the sunlight, to prevent the contraction of the features.

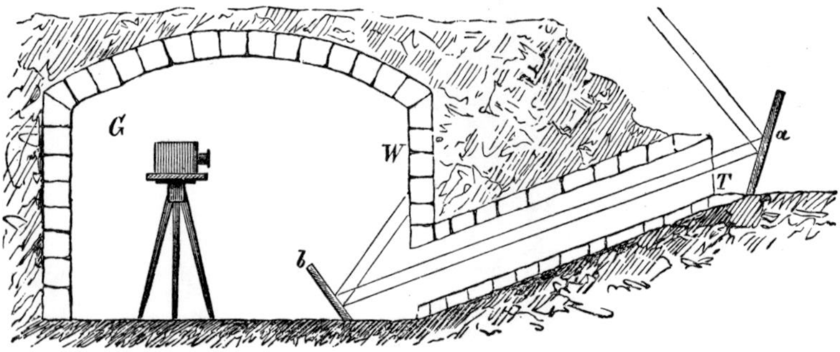

Fig. 24.

It thus appears that all these artificial lights are only auxiliaries to photographic purposes, especially as they are so expensive. Accordingly, their use will be confined to places that cannot be lighted in any other way. The writer of the present work has used sunlight with great advantage when engaged in photographing Egyptian sepulchres. He brought the light into subterranean passages by means of reflection.

Let the reader imagine a mirror set up in the open air, reflecting the sun's rays through the sepulchral entrance *T*, into the subterranean vault *G*. In this vault they are received by a second mirror, which throws the rays on the surface of wall *W*, of which a photograph has to be taken. I admit that nothing but a speck of light is thus received; but if, during the time of exposing the photographic plate, this speck be allowed to move over the part of wall *W*, of which a photograph is to be taken, all parts of the object receive successively enough light to allow of photographic effects. The movement of the speck over the wall is effected by the agitation of mirror *b*.

Braun of Dornach, by help of the same procedure, was able at a later date to reproduce the very dark frescoes of Raphael and Michael Angelo in the Sixtine Chapel and in the galleries of the Vatican, and produced excellent results.

The sunlight remains the most important source of light for photographic purposes. The clearness of this light is, however, exposed to great variations. Even the naked eye recognizes that the sun is much brighter at noon than in the morning and evening. According to the measurements of Bouguer, this difference is so considerable, that the sun at an elevation of 50° above the horizon is 1200 times brighter than at sunrise. The eye, moreover, perceives a decided difference of colour between the sun on the horizon and the sun at the zenith. The latter appears white, the former of a more reddish hue ; and, on making experiments with the spectrum apparatus, it is found that in the setting sun the reddish rays predominate, while the blue and violet are in part wanting.

It follows hence, that the chemical effect of the sunlight is very feeble in the morning and the evening; that it increases as the sun-rises above the horizon, and that it attains its greatest intensity about noon.

The cause of the red hue of the morning and evening sun is found in the fact that the particles of air partly repel the blue rays—for which reason the air (that is, the sky) appears blue—whereas they admit the yellow and red rays more easily.

If *E* (Fig. 25) is the earth surrounded by the atmosphere *A*, *S* the sun at moment of sunrise, *S''* the sun at the moment of sunset for the place *O*, and *S'* the sun at noon, it is apparent that the sun's rays at sunrise and sunset have to travel much farther—namely, the distance between *A* and *O*—than when the sun is at the

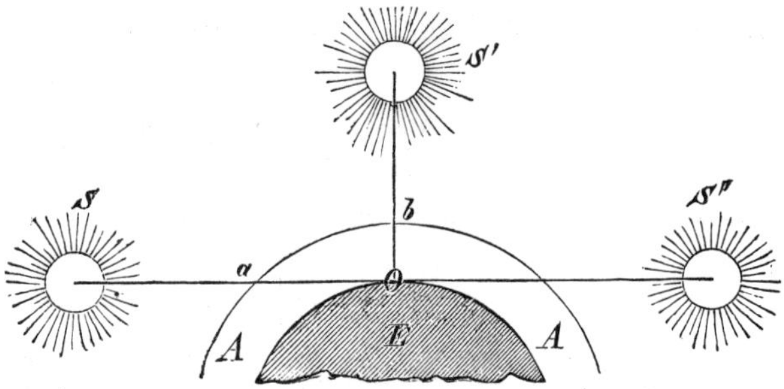

Fig. 25.

zenith *S'*. But in proportion as the stratum of atmosphere through which the sun must pass to arrive at the spectator, the weaker it becomes. It follows from this that on high mountains the chemical effect of the rays of light must be more intense, and this has been proved by experiments on the Alps.

But not only are chemical effects produced by the sunlight; the blue sky, which is nothing but reflected sunlight, is likewise operative, and powerfully so, through its blue colour.

It has been already stated that the blue colour of the sky proceeds from this, that the particles of the air reflect more especially blue light. But the quantity of this reflected blue light varies with the hour of the day, being strongest when the sun is highest (that is, at noon), and it diminishes in proportion as the sun approaches the horizon. Therefore photographers are wont to take their photographs of portraits when they only use the light of the blue sky, at noon; that is, between 10 a.m. and 2 p.m. During these hours the chemical effect of light remains almost the same; afterwards it diminishes rapidly,—quicker in winter, slower in summer. Thus the chemical power of light, according to Bunsen, expressed in degrees, is at Berlin:—

	2 o'clock.	1 o'clock.	2 o'clock.	3 o'clock.	4 o'clock.	5 o'clock.	6 o'clock.	7 o'clock.	8 o'clock.
From June 21	38°	38	38	37	35	30	24	14	6
From Dec. 21	20°	18	15	9	0				

It appears from this example how extraordinarily weak is chemical light in winter (for example, towards noon on the 21st December about half as powerful as towards noon on the 21st June); moreover, how small the amount of chemical light is which is diffused by the blue sky on the 21st December, on account of the shortness of the day. Therefore photographers must expose much longer in winter than in summer, and, their printing process being slower, they take much longer in winter to copy the same number of pictures.

Now the intensity of the blue sky light depends on the position of the sun, and the latter varies, not only according to the different seasons, but also at the very same seasons on different parts of the earth.

If circles be drawn round the earth from pole to pole, we obtain what are called meridians (*m m* Fig. 26). All places that are situated on the same meridian have noon at the same time, but the height of the sun varies very much according to the distance of the place from the equator.

If circles be drawn round the earth parallel to the equator, they form what is called lines of latitude. If the sun is at a particular place on the equator perpendicular at noon, at the 10° of north latitude it is 10° lower; that is, the height of the sun (or the distance of the sun from the horizon expressed in angular measurement) is 80°. At 10° further north, the position of the sun at the same time is only 70°; and at the pole, which is 90° from the equator, the height of the sun = 0; that is, the sun is on the horizon.

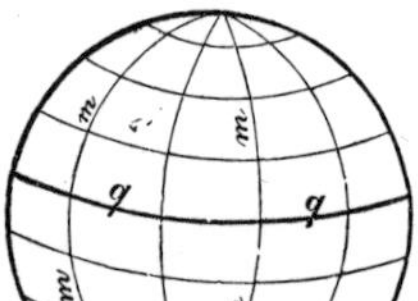

Fig. 26.

The chemical strength of the blue sky light varies greatly, corresponding to the different positions of the sun at the same time. Thus, for example—

At Cairo, on the 21st Sept.,	the strength of light	at noon	=	105°
At Heidelberg	„	„	„ =	57°
At Iceland	„	„	„ =	27°

Therefore, the more southerly a place is, the richer it is in the amount of light it offers to the photographer.

Accordingly, the American photographers are better off than those of Germany and England.

These differences in the intensity of chemical light are yet essentially modified by the state of the weather. If the sky is covered with grey clouds, the chemical intensity of light is considerably less than with a perfectly clear sky. On the other hand, white clouds increase the chemical intensity of light very decidedly. In the autumn the chemical intensity of light is much greater than in spring, perhaps in consequence of the greater transparency of the air. According to Roscoe, it is in August and September more than one and a half times as great.

These variations in the chemical intensity of light are very important to the life of plants. The green leaves of plants inhale carbonic acid and exhale oxygen under the influence of light. But this breathing process does not take place without the presence of light. The green colour of leaves and the variegated scale of colours in flowers only exist under the operation of light. In the dark, plants only develop sickly blossoms, like the well-known white germs of potatoes kept in cellars.

The necessity of light for the life of plants is also seen in the effort made by plants kept in darkened rooms to reach the apertures which admit light, growing as it were towards them. Hence a plant develops with an energy proportioned to the intensity of the light. Accordingly, the greater fruitfulness of the tropics is to be ascribed, not only to the higher temperature, but also to the greater chemical intensity of light. Recent observations have established that the yellow and red rays, and not the blue and violet, produce the greatest chemical effect on the leaves of plants.

We have now arrived at the knowledge of the importance of light for the economy of nature. The atmospheric air consists of two kinds of gas, oxygen and nitrogen, which are in combination. Nitrogen is a perfectly innocuous kind of air, serving to attenuate the oxygen; for the latter alone, though essential to life, would be injurious.

In breathing, part of the oxygen is absorbed in the lungs: it forms, with the organic constituent parts of the body, carbonic acid and water. The carbonic acid and water are exhaled by us and dispersed again in the air.

It is easy to prove by an experiment that a considerable amount of carbonic acid is contained in the air we exhale. Carbonic acid forms, combined with lime-water, an insoluble precipitate called carbonate of lime. If now we exhale through a glass tube, letting our breath pass into the perfectly clear lime-water, the latter becomes troubled by the formation of carbonate of lime. Hence the amount of oxygen in the atmospheric air is continually diminished and converted into carbonic acid. The same result is produced on a larger scale by the process of combustion. In this process a combination takes place of wood or coal with oxygen, and the result is again principally carbonic acid.

It might be supposed from this that, in the course of time, the amount of oxygen in the air must diminish, while that of carbonic acid would increase. This actually takes plåce in closed spaces. Leblanc found that, after a lecture in one of the lecture rooms of the Sorbonne at Paris, the air had lost one per cent. of its oxygen.

In the open air this diminution of oxygen and increase

of carbonic acid gas is not noticed, and the reason of this is that the carbonic acid formed by combustion and the exhalations of animals is again decomposed by plants under the influence of light.

Plants absorb the carbonic acid, retaining the carbon and liberating the oxygen; by which means the latter, lost by combustion and exhalation, is made again available.

There was a time when the atmosphere was much richer in carbonic acid gas than now. When the incandescent and fluid masses that once formed our earth gradually became condensed, when the watery vapours were precipitated as seas, the atmosphere contained almost all the carbon of the earth after combustion; that is, united with oxygen as carbonic acid gas. The air was therefore at that time infinitely richer in carbonic acid than now. When at length the earth had cooled sufficiently for vegetation to be developed, gigantic plants shot forth from the warm ground under the influence of the sunlight. They flourished luxuriantly in the atmosphere rich in carbonic acid, the carbon of the carbonic acid passed over into the form of wood, and thus in the course of thousands of years it was continuously diminished. Revolutions of the earth's surface succeeded; whole territories with their forests were buried under sand and clay beds, and, becoming decomposed, were changed into coal. A fresh vegetation sprouted forth from the newly-formed soil, and again absorbed, under the influence of light, the carbonic acid of the atmosphere, to be once more engulphed by a fresh cataclysm. Thus, the carbon from the carbonic acid of the atmosphere was stored as coal in the depths of the earth; and thus the atmosphere, by the chemical effect of light, became continually richer in oxygen, until at length, after countless revolutions of the earth, it obtained that wealth of oxygen which made the existence of man possible, when he appeared at the end of the earth's development.

We see, therefore, that the chemical influence of light has played an important part in the development of our planet, and it continues to do so in the economy of nature.

CHAPTER IX.

ON THE REFRACTION OF LIGHT.

Simple Refraction—Deviation—Index of Refraction—Refraction in Glass Plates—Prisms and Lenses—Production of Prisms or Images by Lenses.

We have already pointed out (p. 60) that when a ray of light passes the border of two transparent media of unequal density, a change of direction takes place which is called refraction.

If a small coin is placed in an opaque vessel, and the eye *O* be kept in such a position that the edge of the vessel covers the coin, it is invisible. But if water be poured into the vessel the coin becomes visible, and this takes place by the refraction which the rays experience in passing from water to air. (See Fig. 27.)

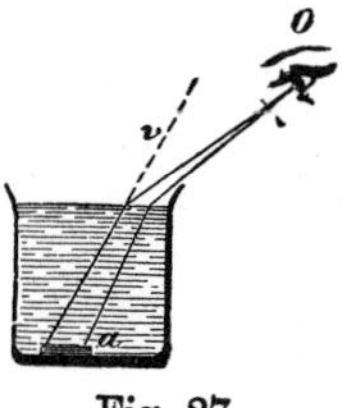

Fig. 27.

The angle which the united rays make, before and after the refraction, is called the deviation.

This deviation increases in proportion to the oblique-ness with which the rays fall upon the surface of the water.

In order to determine exactly the degree of the refraction, let a perpendicular line be conceived to be erected at the point of immersion *n* of the ray *n l* (Fig. 28). This line is called the normal, or plumb line, and the angle *i* which the ray forms with this normal is called the angle of incidence, while the angle *r* which the refraction ray forms with the same normal is called the angle of refraction.

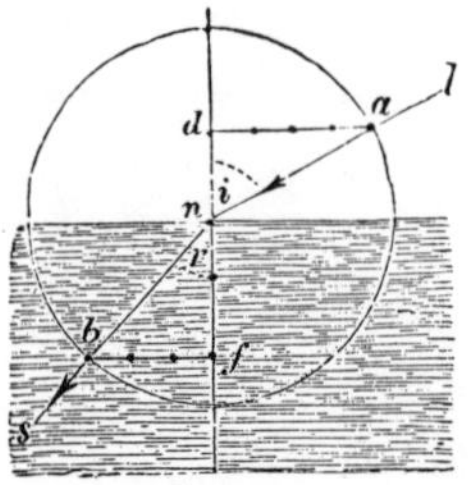

Fig. 28.

The ratio of the magnitude of the angle of incidence to the angle of refraction is peculiar. If a circle be described, and from points *a* and *b* perpendicular lines *a d* and *b f* are let fall on the normal, the result obtained is what mathematicians call the sine of an angle. Thus *a d* is the sine of *i*, and *b f* the sine of *r*. The ratio of the sine of incidence to the sine of refraction is constant.

This ratio is when light leaves air for water as 4 to 3; that is, the sine *b f* is $\frac{3}{4}$ times as great as sine *a d*, or sine *a d* is $\frac{3}{4}$ times greater than sine *b f*. Light is still more refracted on entering glass. In this case the ratio of the sines is as 3 to 2. This ratio of the sines of the two angles is designated by the name exponent of refraction, or index of refraction.

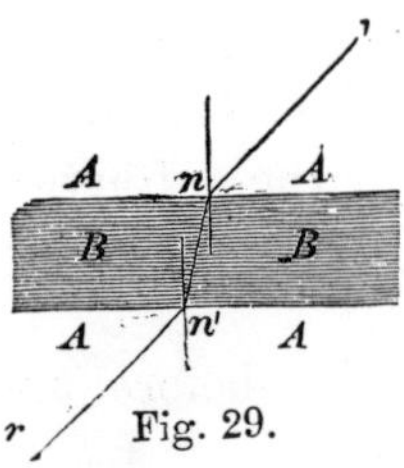

Fig. 29.

If a ray of light *n l* falls upon a smooth sheet of glass, it experiences a similar refraction; it continues in the direction *n n*, and the angle of refraction at *n* on the glass becomes two-thirds of the angle of incidence. (Fig. 29.)

On issuing from the other side of the sheet of glass, another refraction takes place; but in this case the angle of refraction at *n′* in the air becomes one and a half times larger than the angle on the glass, and as the angle at *n* is equal to the angle at *n′*, the angle of emergence *r n′* is of the same magnitude as the angle of immersion *n l*; that is, the ray continues, after refraction, in its original direction. At all events, it only experiences a prolongation parallel with itself. Therefore we see through our windows in the same direction in which they are really situated.

The ratio is entirely different when the spectator looks through a glass having three faces. If the eye is at *o*, and an object at *a*, and a prism with three faces be held close to the eye, the object is not seen at *a*, but in the direction of *a′*. The incident ray *a d* suffers a deviation at the first face of the glass, taking the direction *d c*; at the refraction on the second face it makes another, *o c*. Both deviations correspond.

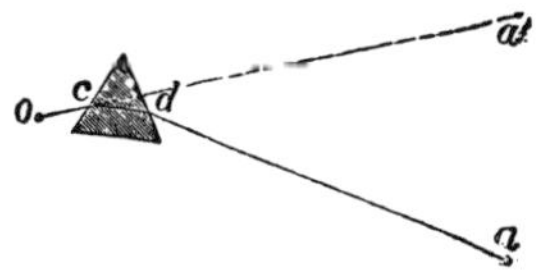

Fig. 30.

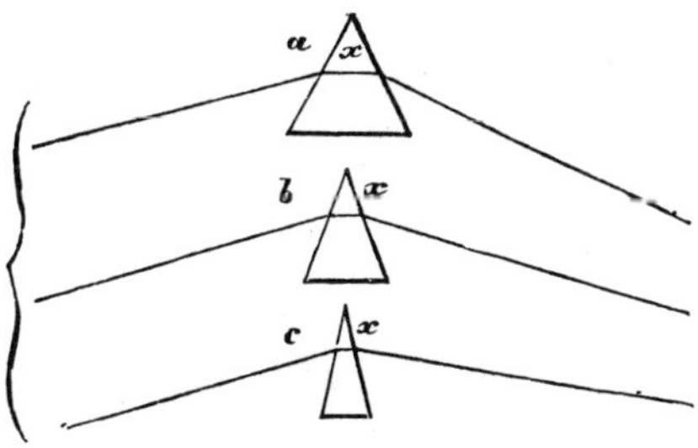

Fig. 31.

The greater the magnitude of the angle *x* which the two faces of the prism, through which the ray passes, make with each other, the greater is this deviation.

5

Thus the deviation at prism *d* is greater than at prism *c*, and at prism *a* it is greater than at prism *b*; because the angle of refraction *x* in *b* is greater than in *c*, and at *a* it is greater than in *b*.

If a glass structure be erected, consisting of separate prisms of different angles, and if a bundle of parallel rays be conceived to fall upon it, the ray *a* is more

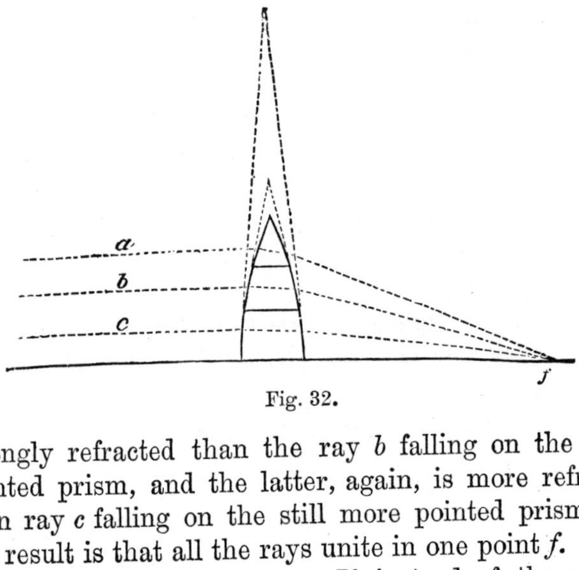

Fig. 32.

strongly refracted than the ray *b* falling on the more pointed prism, and the latter, again, is more refracted than ray *c* falling on the still more pointed prism, and the result is that all the rays unite in one point *f*.

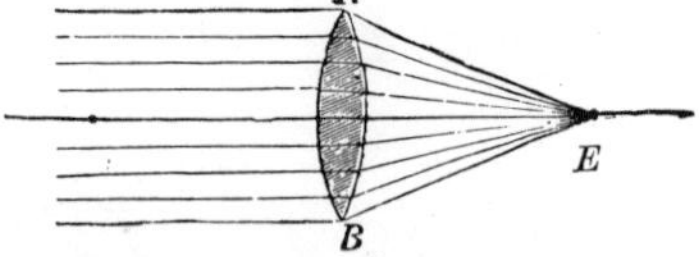

Fig. 33.

If instead of the prisms we substitute a connected symmetrical mass of glass, we obtain the section of a burning glass, or, as the opticians say, a lens, which has the property, as in the illustration, of uniting all parallel incident rays in one point. (See Fig. 33.)

Every lens is contained between two spherical faces. The connecting line running through the centre of both spherical surfaces is named the axis of the lens, and point *E* (Fig. 33), where the parallel incident rays unite, is the focus, while its distance from the lens is the focal distance. But not only are the parallel incident rays united in one point by the refraction in a lens of this kind,—the same thing occurs with all rays issuing from the same point. Their converging point is named principal focus.

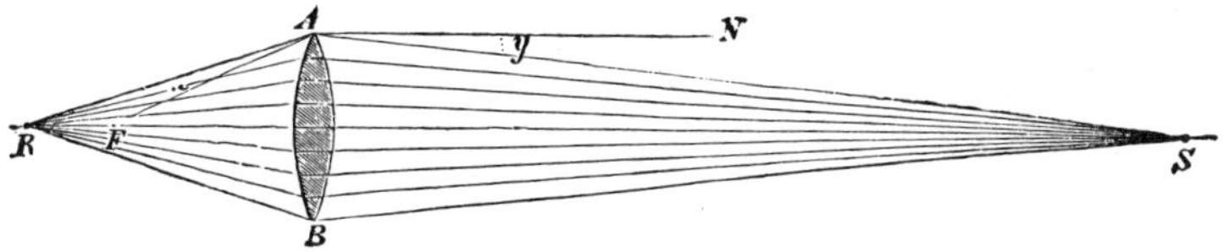

Fig. 34.

A luminous point *S*, for example, directs a cone of rays to the lens. After refraction these are united at *R*. If *S* be brought nearer to the lens, *R* removes further; if *S* be brought so near that its distance from the lens is twice the focal distance, then the converging point *R* is equally distant from the lens.

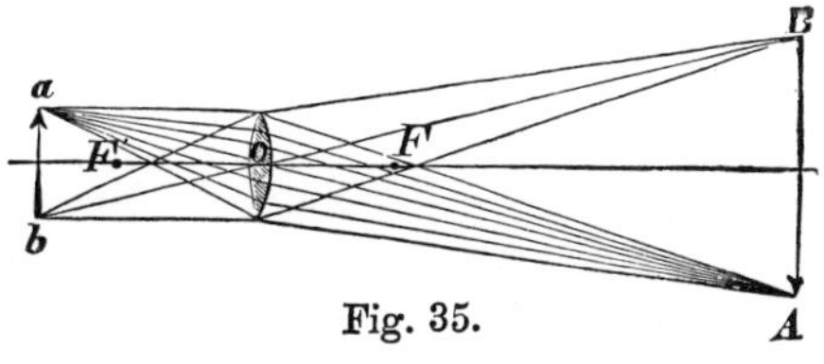

Fig. 35.

If instead of the luminous point, an object, for example an arrow *B A*, is placed before the lens, each individual point of the object sends out a cone of light to the lens, and all the rays of one and the same cone converge to one point, the rays issuing from *A* to *a*, and those issuing from *B* to *b*; and the result is that a perfect miniature and inverted image of the arrow is produced.

If the arrow be moved nearer to the lens, its picture is removed farther from the lens and is magnified. For example, if the little arrow *a b* is placed before the lens, it produces the enlarged picture *B A*.

But if the arrow be removed farther from the lens, its image approaches continually nearer to the lens, and therefore becomes continually smaller. Accordingly, a lens is able to project enlarged or diminished images of an object, by the latter being approached to or removed from it.

CHAPTER X.

THE PHOTOGRAPHIC OPTICAL APPARATUS.

Construction of the Camera Obscura—Telescopic Images—The Magic Lantern—Magnifying Apparatus—The Stereoscope.

We have just shown that a lens is able to produce enlarged and diminished pictures of objects according to their distance. On this principle depends the working of

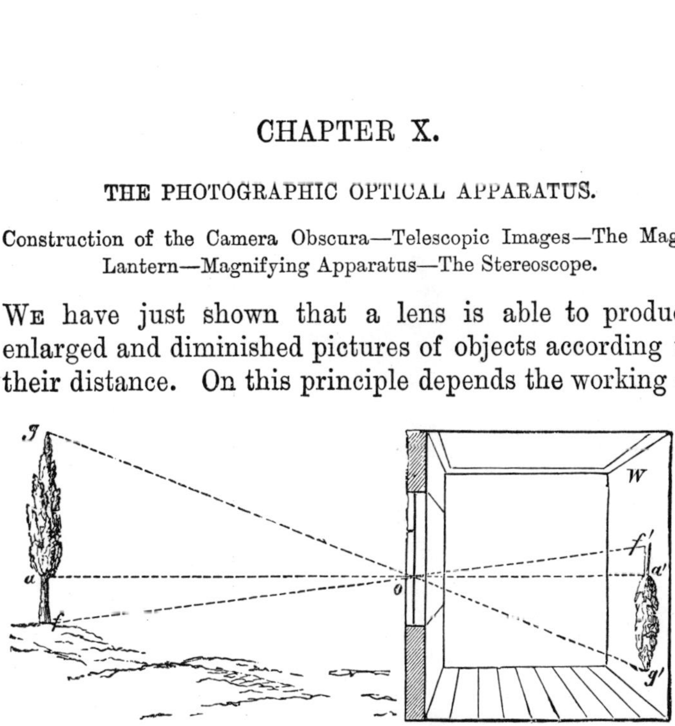

Fig. 36.

the camera obscura, the most important photographic apparatus, which serves to project plane pictures of solid objects in nature. We have already described (see p. 7) its simplest form. It is a dark chamber having a small hole in its lid. This arrangement produces very indistinct pictures deficient in light. But if a lens is placed in the dark slide *o* (Fig. 36), this produces on the opposite wall a picture of the objects facing the chamber, which is much clearer and more sharply defined than the image produced through the aperture. It is evident that in this case the distance of the wall must correspond to the distance of the image. Now, as this varies, in order to determine the place where the image is found the camera has been converted into a small dark box (Fig. 37), the back part of which is movable, and contains the ground glass slide *g*. If the back of the camera *o* is moved to and fro, it is easy to find the situation of an image of an object placed before the lens *l*. In order to ascertain this distance with the necessary accuracy, photographic lenses have been supplied with a screw having a spring *r* at the frame of the lens; but this addition is by no means necessary.

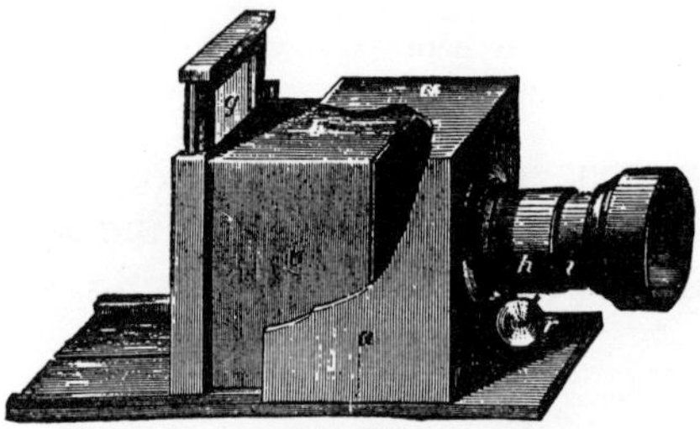

Fig. 37.

In order to be able to see on the ground glass slide *g*, all foreign light that blinds the eye must be kept off, and to this end a dark cloth is thrown over the head, forming what is called the black curtain.

The operation of seeking the image is named in photography focussing. It follows from what has been said that the image appears inverted on the ground glass slide. Though the process of seeking the image appears simple at first sight, it is really rendered difficult by the fact that objects at different distances present images that likewise vary in distance from the ground glass slide. For example, if a head be placed opposite a camera, the nose is nearer to the lens than the hairs behind the head; and the result is that the image of the nose in the camera is farther from the lens than the hairs of the back or side of the head. Accordingly, the whole head never presents the same degree of sharpness or definite accuracy. Photographers are satisfied with rendering the main points with definite clearness, such as the face, and they bestow less care on the subordinate parts.

If the situation of the object be rather remote (for example, a landscape, whose nearest features in the foreground are about fifty times as remote as the focus), the images of the different features, whatever their remoteness may be, appear all in the same focus.

The same thing occurs in the case of stars. Photographic cameras are well adapted to project images of stars, only they are very small if the focus of the lens is small. Accordingly, telescopic lenses are preferred in such cases. The production of images in these cases rests on the very same principles as the production of images in other lenses. If we imagine the telescopic lens *o o*, and place

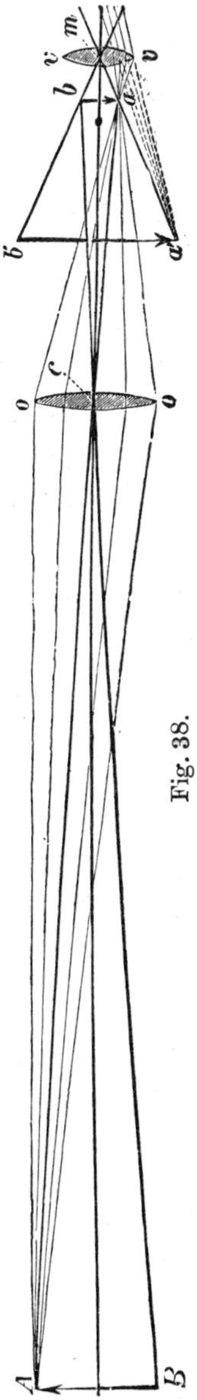

Fig. 38.

before it at a great distance the arrow *A B*, it produces a very diminished image of the arrow *b a*. Thus, the image of the sun at a distance of 90 millions of miles is, in the focus of a lens six feet wide, only eight lines in size. If it is wished to obtain the photograph of such an image, the tube *R*, on which the lens *L* is fixed, must be arranged as a photographic camera. (See Fig. 39.) A movable ground glass shade *n* is brought up behind, to throw out a sharply-defined image, and this can be exchanged for a photographic plate during the exposure. This is the method adopted by Warren de la Rue, Rutherford, and others engaged in expeditions to determine eclipses, and also by the author in the expedition to Aden in 1868.

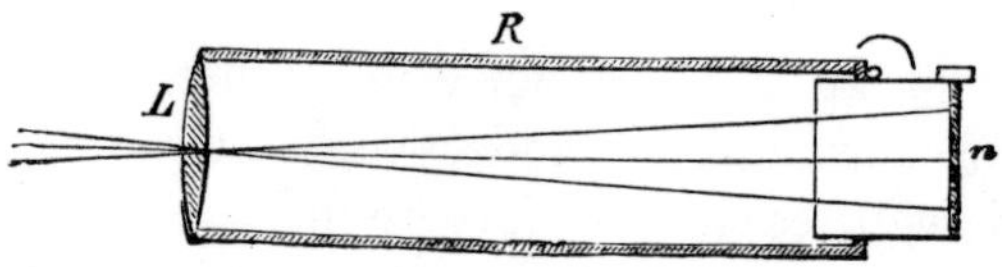

Fig. 39.

The images taken by a photographer are usually smaller than nature. But he is also in a position to produce images that are larger than the originals. Every lens gives, as described at p. 88, different images of the same object, varying according to distance. If the object is nearer than twice the focal distance, an enlarged image is produced, but if it is more remote, the image is smaller. The latter is the commoner case. Nevertheless, to make enlarged images direct from nature is attended with difficulties. The larger the image, the larger is the surface over which the light is dispersed which issues from the objects, and, accordingly, the smaller will be the amount of light over each part of the image. But, in proportion as an image is deficient in light, the larger must be the exposure to produce a photographic impression. A man would find it difficult to endure such a long sitting, therefore this method is only employed in drawings and the like.

Enlarged pictures of other objects are represented by the help of an apparatus resembling the magic lantern. The magic lantern consists in the production of an enlarged image by means of lenses. Instead of a simple lens, a system of lenses *n n o o* is employed for enlarging, and gives more sharply defined images. The image that is painted or photographed on glass is slid in by a lateral slide and brightly illuminated. To this end a lamp *L*, the concave mirror *H*, and the lens *m m* are employed. These concentrate the clear lamplight on the image that has to be enlarged. According as lens *n o* is approached or withdrawn, — that is, according to the variation of its distance from its original,—the images obtained are larger or smaller in size.

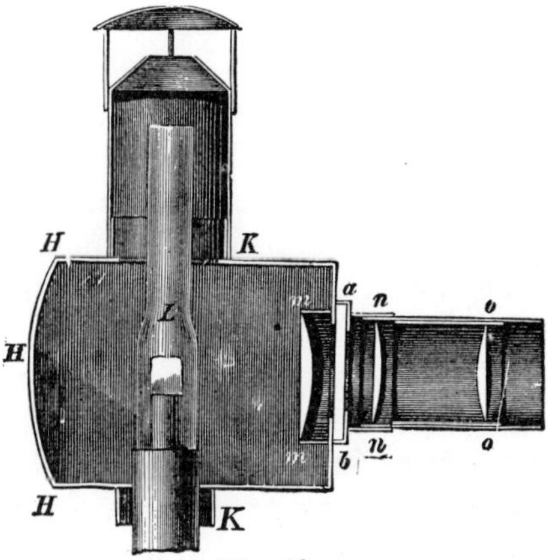

Fig. 40.

This instrument was formerly nothing but a plaything, but it has become latterly an important auxiliary in instruction. Photographs of microscopic preparations, of animals, plants, minerals, landscapes, national types and architecture, give in this manner a more faithful and a truer representation than the maps and tables, which are in general very imperfectly designed.

In America this application of the magic lantern is universal. Every considerable educational establishment possesses one such instrument, and often more. In Germany, though so useful, it has been hitherto left in the hands of the traders connected with the annual fairs, who employ them for what are called cloud pictures. These cloud pictures are produced by the help of two magic lanterns placed side by side, both of which project their images on the same screen. If one of these lenses be covered up, one of the pictures disappears and the other alone remains visible. Meanwhile, if another image be substituted for the one withdrawn, and the lid again taken off, a combination of two images is obtained. If the lens be only gradually, not suddenly, closed, the image also fades away gradually, until it disappears entirely.

Professor Czarmak, at Leipsic, has latterly introduced the representation of enlarged images by the magic lantern as an important auxiliary in his lectures; and he obtained so great a success with it that he prepared the way for its general introduction into schools.

We take this occasion to remark that wonderful paintings on glass have been lately produced from photographs by means of a new light-printing process. These glass images have been exposed for sale, and are specially destined for the magic lantern. Their price is so moderate, and the objects they represent are so interesting—being landscapes of all parts of the earth—

that it is within the reach of every family to obtain a collection of the most pleasing and entertaining views. At domestic entertainments by the family fireside, images of this kind united with the magic lantern become an important means of instruction and enjoyment to young and old.

A petroleum lamp is not sufficient for the representation of such images on a large scale. For this purpose more powerful applications of light must be employed,

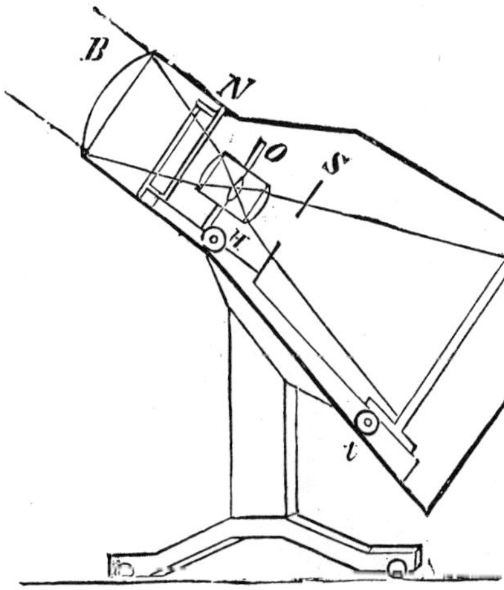

Fig. 41.

either Drummond's lime light or the electric light (see p. 72). To retain such enlarged images in a photographic form, a sheet of sensitized paper is stretched over the place and instead of the screen.

To produce photographic images life size, the magic lantern is not used, but the solar camera, a section of which is represented Fig. 41, and a front view of it Fig. 42.

Sunlight is suffered to fall on a great lens *B*, which oncentrates it on a small negative *N*, and close to it is the objective *O*, which projects an enlarged image on the screen *R*. The image is evidently negative. If a sensitized piece of paper be stretched at *R*, this paper be-

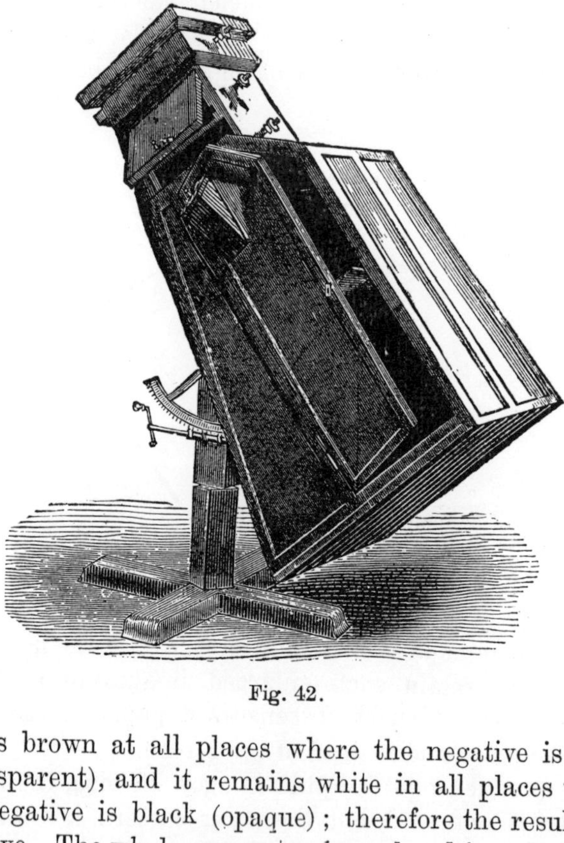

Fig. 42.

comes brown at all places where the negative is clear (transparent), and it remains white in all places where the negative is black (opaque); therefore the result is a positive. The whole apparatus is enclosed in a darkened wooden box, which can be shifted by means of cog

PLATE II.

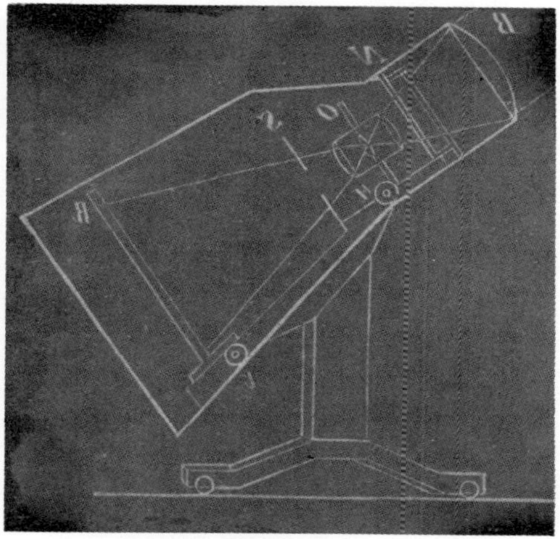

Negative.

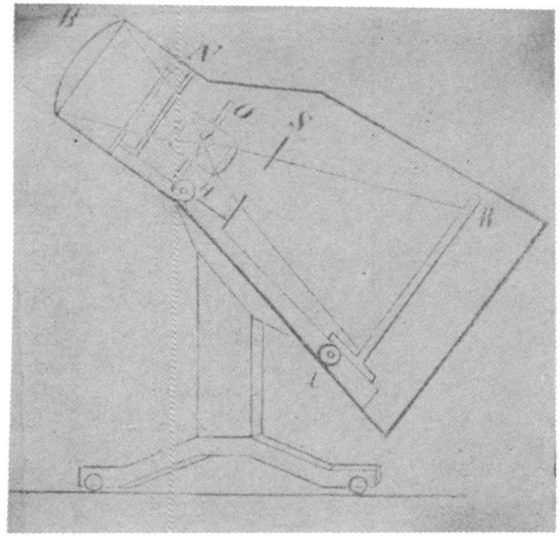

Positive.

LICHTPAUS PROCESS.

See p. 96.

wheels and a winch, so that it can always be turned towards the sun.

In conclusion, we have to describe a most beautiful optic-photographic apparatus, which permits us to see images not only as plane objects, but as solid bodies. This is the stereoscope.

Our readers know already that this instrument is intended to exhibit double pictures, the two halves of which on the first look seem to be absolutely alike, and which when viewed through the instrument form one picture, which appears no longer plane, but solid.

The two pictures which are seemingly alike are, in fact, different. If we look at one of the cubes with the right eye, we see rather more of the right side; if we look with the left eye, we see something more of the left side, taking for granted that the head is not moved. The pictures of the right and of the left eyes combine with each other and give the solid appearance or impression.

If we close one eye, the bodily impression is far weaker; the objects appear plane or flat. This may not be readily credited, because men do not often seek an explanation of what they see, but look at objects much too hastily. But it can be easily ascertained that such is the fact if a bottle be placed before a wall, or in front of an upright book, and if we then look at it. With two eyes open we readily perceive the distance of the bottle from the wall or book, but directly we close one eye the bottle and the book appear to be almost contiguous, and it is only by moving the head on one side that we clearly distinguish the distance between them.

Accordingly, the use of both eyes is necessary to have a proper perception of a bodily impression. It is only in this manner that we come to the conviction that space has not only height and breadth, but also depth. One-eyed persons only receive this impression by turning the head on one side. If objects are very remote, the difference between the views which the right eye and the left eye have of them is very inconsiderable; and, accordingly, such remote objects appear flat and without solidity, and it is only when we change our position and observe it from different sides that we become acquainted with its solidity. This is therefore purely an affair of experience. Every person will recognize a remote house as a solid object, because we know from experience that a house is a solid; but that we actually see it as a flat surface is proved by the deception produced by theatrical decorations, where the remote background, if properly painted, often produces an extremely natural effect; but we perceive this background to be flat directly we move the head on one side. When this is done, a solid object presents a different appearance, but a flat surface remains unchanged.

Wheatstone was impressed by the fact that the solid impression made by an object is the combination of different representations of it given by the right and the left eye. Accordingly, he tried to substitute for one object a picture of the right side of it for the right eye, and a picture of the left side of it for the left eye. He obtained in this manner a perfectly solid impression, though the double picture occasioning it is no solid at all. Nevertheless, some people are able to see stereoscopic pictures as solids without using an instrument.

But most persons require an apparatus which renders it possible for both eyes to see in the same place the two pictures that are separated by a certain distance. This apparatus is the stereoscope. Its most essential features, as seen in the accompanying diagram, are the double picture on the slide, the partition on the interior of the box which prevents the right eye seeing the picture on the left, and *vice versâ*. Further, there is the lid, which is generally provided with a mirror which can be either shut or opened, so as to exclude or let in the light into the box, and lastly the two eye-glasses in the front.

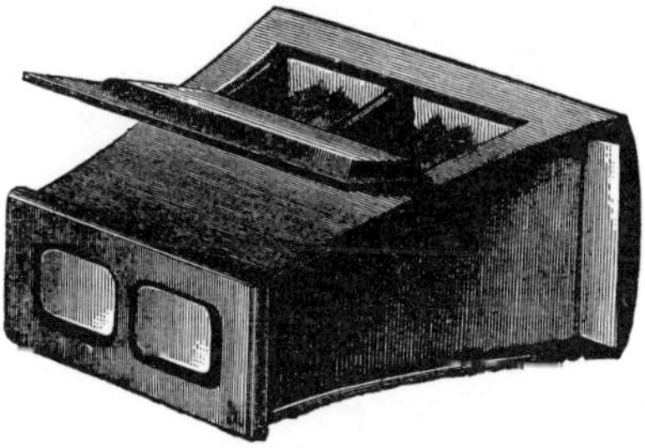

Fig. 43.

These eye-glasses are represented in the diagram as adjacent; they are two halves of one lens, and work in the same manner. (Fig. 44.) We are indebted to Brewster for the construction of this instrument.

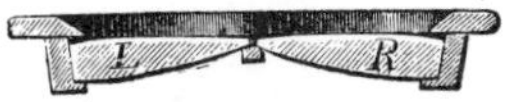

Fig. 44.

We have shown at a previous page that a lens gives a diminished image of a remote object, and an enlarged inverted image of a near object. This image is objective, that is, it can be clearly discerned on the ground glass shade of a camera. Nevertheless, this phenomenon only takes place when the object is more remote than the focus. The case is different when the object is nearer to the lens. Let an ordinary magnifying or burning glass be held near some writing, and it will be seen upright, and not inverted. The image appears also enlarged, but on the same side as the object, and the accompanying diagram illustrates the manner in which it originates. Thus *F* represents the focus of the lens, *A B* an object within the focal distance, and *a b* its image, as it appears to the eye upon the other side of the lens.

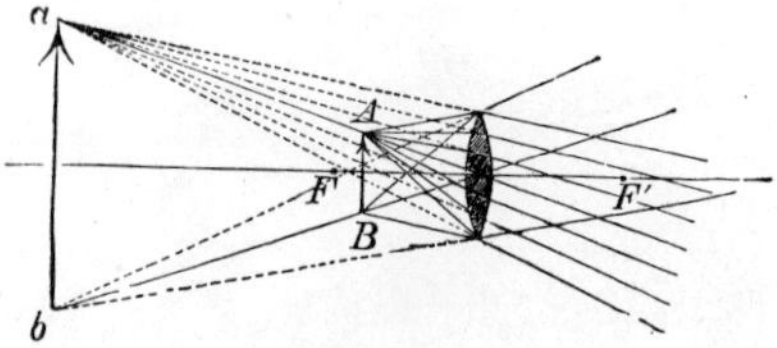

Fig. 45.

As may be seen in the diagram, the rays issuing from *A B* do not actually unite in one image, but their directions are thrown back in the pointed lines of the diagram and lengthened till they unite in one point, and there we see the image. The eye naturally seeks the object sought for always in the direction of the incident rays, as may be seen, for example, in a mirror, when we see the mirrored objects behind it.

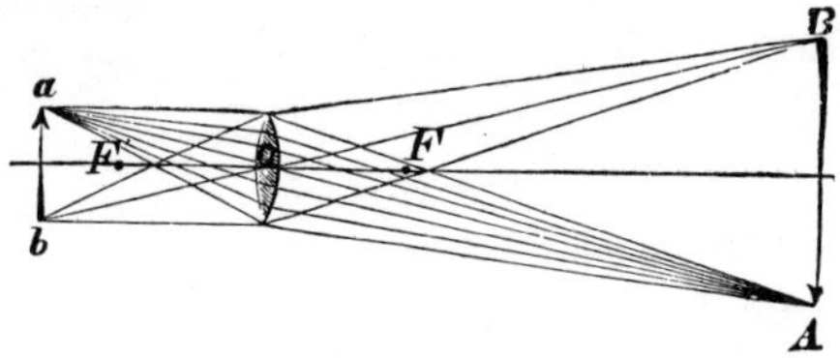

Fig. 46.

In order to give a clear illustration of the working of a lens, for near and remote objects, we introduce the same diagram of p. 87 again, which shows the production of an inverted and enlarged image *B A*, representing the arrow *a b* situated within the focus.

A lens employed to see objects enlarged within the

focus is named a microscope, magnifying glass, or loupe. These magnifying glasses are the lenses of a stereoscope. They present us with a rather enlarged upright image of the object seen, but they produce likewise the effect of a prism. As may be seen from Fig. 44, the two lenses consist properly of only two half lenses, which are joined in an inverted direction, thus making the impression of prismatic glasses, and working similar effects.

We pointed out at a previous page that an eye *o* sees an object *a* through a prism in the direction *a a′*, that is, raised to the upper side of the angle of refraction of the prism. The same thing happens with stereoscopic glasses. We see the image, not in the original direction, but deflected to the side of the angle of refraction, that is, through the centre of the instrument.

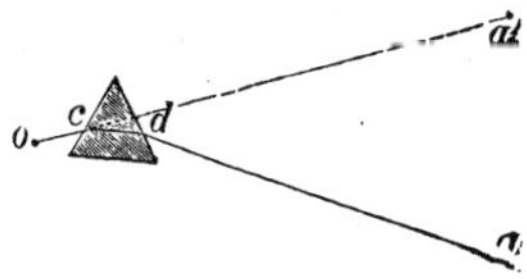

Fig. 47.

The two corresponding points *a a′*, which belong to the right and left images, appear therefore in common to both eyes at *a″*—that is, at the same place—and consequently both eyes see only one image instead of two.

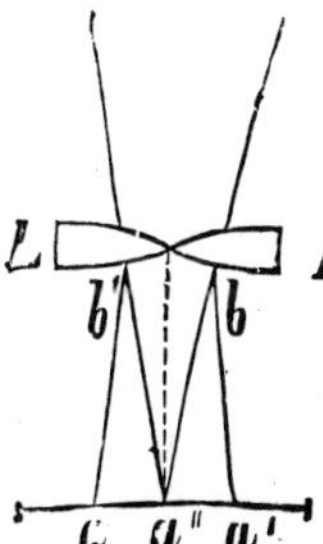

Fig. 48.

Now, it is proper to remark that every one who wishes to see an object (for instance, writing) clearly and distinctly holds it at a definite distance from his eye. This distance is the distance of clear vision. In the case of good eyesight it is eight inches; with far-sighted persons it is more, and with short-sighted persons it is less. In the case of stereoscopic vision, the image appears remoter or nearer, according as it is removed from or approached towards the two glasses. If the image is near the two glasses, it appears when viewed through them nearer and smaller. In the opposite case it appears farther and larger. But every one wishes to see the image at the distance of clear vision, therefore stereoscopists must have glasses that can be shifted, in order that persons may adapt the position of the image to the eye; that is, that they may vary the distance of image and glass until the image shows in the clearest manner to them. If apparatus for this purpose is wanting, the instrument is not adapted for eyes of average power of vision, and requires an effort in eyes of a different calibre. Persons are often met with whose eyes are not of equal strength, one being short and the other far-sighted. There can be no satisfactory stereoscope for such cases; for if the distance of the lenses is adapted for one eye, it does not suit the other.

Nevertheless, such persons can obtain a tolerable stereoscopic vision if they hold a suitable eyeglass before one eye.

A great hindrance to viewing stereoscopic pictures on paper is the shape of the Brewster stereoscopic box, which is closed all round and only open at the top. This aperture only admits an insufficient amount of light to the picture, which is commonly left in the shade on one side.

This defect has been removed in the American stereoscope, which dispenses with any box. Glasses are fitted in a frame *g g*, which is held firm by means of a handle; the partition *b* serves to separate the field of view of both lenses. The image is placed on the crossboard with the wirework *d d*, and this board can be easily slid to and fro, so that the proper position of the image with reference to the eye may be found.

But the American stereoscope is only suitable for paper pictures. The beautiful transparent stereoscopic pictures on glass can, however, only be viewed with Brewster's stereoscope, as they must be seen in light from above, and all light reaching them from the front must be excluded, or the effect will be destroyed. At Berlin, the firm of Moser and Company prepares stereoscopes on the American plan.

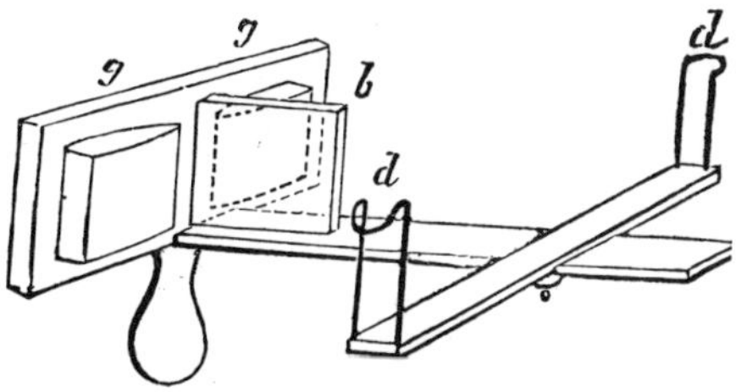

Fig. 49.

We have named the stereoscope an optic-photographic apparatus. But we remark that double pictures drawn by the hand can also be viewed through it. It is evident that the projection of such pictures only succeeds in very simple subjects. It would be very difficult to represent stereoscopically a complicated object; for example, a man, a landscape, or a machine. This was only possible by means of photography, which can give with the greatest ease a pictorial reproduction of the most complicated objects from any point preferred. It is only since the invention of photography that the stereoscope, which was formerly a piece of furniture in collections of philosophical instruments, has become a favourite instrument with the public. Notwithstanding their small form, the pictures of these instruments make a clearer and more intelligible impression than pictures of the same object in a larger form. A single picture of a machine, or of complicated architecture (for example, the choir of the Cologne Cathedral), is often a hopeless maze of details. But in the stereoscope the confused masses are directly defined; they become distinct in perspective, and the eye perceives with great clearness the interior structure. In this respect the stereoscopic pictures are of equal value to the magic lantern in imparting instruction.

CHAPTER XI.

THE CHEMICAL EFFECTS OF LIGHT.

Physical and Chemical Processes—Moser's Experiments—Effects of Light upon the Elements—The Manner in which Phosphorus, Oxygen, and Chlorine act in the Light—Effect of Light upon Salts of Silver—Effects of Light on Chloride of Silver, Bromide of Silver, and Iodide of Silver—Theory of Development—Dry Plates—Theory of the Positive Process.

In the previous chapter we have become acquainted with the part which light plays in photographical processes. We will now enter into the domain of chemistry, in order to explain the phenomena which occur in consequence of the irradiation of light on substances sensitive to its influence.

All bodies in nature are perpetually subject to changes. The sun, moon, and stars change place; wood and sugar can be rubbed into dust, and change form; lead can be melted, and thus altered in its state of aggregation. These modifications, which we have instanced as illustrations, do not affect the substance of the bodies thus altered. Wood may be rubbed or sawn into the finest dust, yet it remains wood; lead remains lead, notwithstanding the melting process. Changes of this kind, that leave the substance of bodies unchanged, are styled *physical changes*.

But, besides these modifications, there exist others of a different nature. If a piece of wood be heated in flame, it is consumed; in this case the nature of wood is completely destroyed. It becomes changed into combustible gas, becomes cinder and ashes, slack and friable, and leaves nothing but a heap of ashes instead of wood. A rod of iron, brought to a red heat in the air, becomes pliant and coated with a black crust, which falls off in powder when struck by a hammer, constituting the black or magnetic oxide. In this case the substance of the iron is totally changed. Changes of this kind are styled *chemical changes.*

Now, light is able to produce chemical as well as physical changes. We have already stated that the mineral styled red sulphuret of arsenic is decomposed into a yellow powder when exposed to the light. It can be melted, in which case it forms, on cooling, compact red masses, which are again decomposed on a further exposure to the light. The number of physical changes in this branch occasioned by light is not great, but the phenomena are in themselves remarkable.

Moser has remarked that light operates on almost all surfaces. He covered smoothly-polished surfaces of silver, ivory, and glass with a friable coating, and exposed them to the light. After this he fumigated it with vapour of mercury, and found that the fumigation or vapour was condensed most powerfully where the light had operated on the surface. Accordingly, Moser established the proposition: Light works on all bodies, and its operation can be made visible by vapours, which are condensed on the parts exposed to light.

The chemical changes effected by light are far more

numerous than the physical, and their study is the special problem of photo-chemistry.

Before passing to complicated phenomena in photography, we must make the reader acquainted with the simpler phemonena of the art.

(*a*) *Operation of Light on the Elements.*

The chemist understands by the term elements simple insoluble bodies. Thus water, which the ancients named an element, is no element in the chemical sense of the term, for it can be easily decomposed into two components of a gaseous nature—oxygen and hydrogen. Air, also an element of the ancients, is no element viewed in the light of chemistry, for it is a combination of two kinds of air—oxygen and nitrogen. But the two latter bodies, oxygen and nitrogen, are undecomposable bodies or elements. The chemical elements are the well-known metals; also sulphur, phosphorus, chlorine (a greenish, strong-smelling gas developed from chloride of lime); further, the less known substance bromine (a brown, unpleasantly-smelling substance of a fluid nature); lastly, iodine (a black substance, also of a fluid nature, and used for friction). All these elements unite together and produce bodies with new properties. Metal iron unites with the light oxygen, and produces the red powdery iron rust. Sulphur unites with oxygen, and produces the pungent, strong-smelling sulphuric acid. Iodine and chlorine unite closely with metals forming the metallic iodides and chlorides, which have quite peculiar properties. Amongst these are iodide of silver and chloride of silver.

It is remarkable that many elements present themselves in quite different states, so that it might be supposed they were quite different substances. The yellow, inflammable, poisonous phosphorus, soluble in ether, and formerly used in the manufacture of matches, is changed by heating in a closed vessel into a reddish substance difficult to kindle, not poisonous, and insoluble. This is, however, simply phosphorus, and passes by melting into the state of common phosphorus.

It is an interesting fact that this transformation of yellow into red phosphorus is effected not only by heat, but also by light. If yellow phosphorus be exposed for a long time to the light, it becomes red.

The oxygen of the air is also susceptible of similar changes. The ordinary oxygen is a colourless and inodorous mass. By the operation of electricity, however, it is easily changed into another kind of gas, distinguished by a peculiar smell—the sulphurous smell attending lightning when striking the earth. This new gas has a much more oxidizing or rusting effect than common oxygen, and oxygen thus transformed is named ozone.

This ozone is also formed by the operation of light: if oil of turpentine be poured into a large bottle containing air, and if it be agitated violently in the sunlight.

Equally peculiar are the changes experienced in the sunlight by two other elements not so well known, chlorine and bromine, which have only been carefully observed latterly.

Chlorine is a yellowish-green gas, with a disagreeable smell, developed by fumigating with chloride of lime, and distinguished by its properties of destroying or bleaching coloured stuffs, and annihilating infectious matters. Bromine is a body very similar to it, but in a fluid not in a gaseous state at an ordinary temperature, though it can be easily vapourized, and then appears as a brownish-red gas.

Chloride and bromide gas show a peculiar relation to light, even in their combinations.

Chlorine gas has a very peculiar relation to hydrogen — a gas which forms the chief constituent of water, and can be easily obtained from it, if zinc be thrown in and diluted sulphuric acid added. When this is done, the zinc attracts the oxygen of the water, forming, with the sulphuric acid, sulphate of zinc, while the hydrogen escapes in the form of gas.

If this combustible kind of gas is mixed with chlorine gas, and the mixture is exposed to the sunlight, an explosion takes place. This accompanies the chemical combination of chlorine gas and hydrogen in a new body —*hydrochloric acid*—having no resemblance to chlorine or to hydrogen. This acid is of a sour taste, very soluble in water, does not bleach like chlorine, and is not combustible.

Another body—iodine—is very closely related to chlorine and bromine. Iodine is a solid body, appearing in the form of shining black crystals, and emitting when heated a wonderful violet vapour.

(*b*) *Chemical Effect of Light on Salts of Silver.*

Iodine and bromine unite with metals like chlorine, forming the iodides, bromides, and chlorides of metals. Kitchen salt is one of the commonest combinations of this kind, consisting of chlorine and sodium. Sodium is a metal not employed in the industrial arts, very powerfully attracting the oxygen of the air, forming rust, so that it has to be protected by being kept under naphtha. The chlorides, bromides, and iodides of the metals all show a nature analogous to salt. Chloride, bromide, and iodide of silver are particularly interesting to us. These three salts are obtained if we allow chlorine, bromine, and iodine to operate directly on silver; but a more rapid method is to dissolve in water chloride of sodium, bromide of sodium, and iodide of sodium, and to add to them a solution of the salts of silver.

Silver also forms salts in combination. If a silver coin is thrown into nitric acid, it is dissolved, forming nitrate of silver; and this is obtained after evaporation as a white soluble salt.

If a solution of this be mixed with a solution of chloride of sodium, a white cheesy precipitate of chloride of silver is obtained by both salts interchanging their constituent parts. Chloride of sodium and nitrate of silver produce chloride of silver and nitrate of sodium.

Bromide of silver is produced exactly in the same manner if bromide of sodium be added to a solution of silver, and iodide of silver is produced by adding a solution of iodide of sodium also to a solution of silver.

Bromide of silver and iodide of silver are thus separated as cheesy precipitates, because they are all three insoluble in water. If they are saturated by being placed upon filtering paper, having water poured upon them, after being dried chloride of silver produces a white powder, bromide of silver a yellowish white, and iodide of silver a yellow powder. All three are very tenacious bodies, not easily decomposed by heat, nor soluble in water, alcohol, or ether, but they can be dissolved in a solution of hypo-sulphite of soda and cyanide of potassium, by combining with these bodies to form new chemical compounds which are soluble in water.

These three combinations,—chloride, bromide, and iodide of silver,—which are peculiarly tenacious bodies, show a marked sensitiveness to light, and this sensitiveness is the basis of modern photography.

By the light of a gas lamp in a dark room the chloride of silver appears perfectly white, but it quickly takes a violet tint in the daylight. It is often said that it becomes black; this, however, is not the case. This violet colouration is the result of a chemical decomposition. The chloride is liberated and disappears partly as a greenish gas, which, from its abundance as well as its odour, can be perceived to be chloride of silver. The violet powder which remains behind at an earlier period would have been taken for metallic silver.

Metallic silver can, of course, under certain circumstances, present itself in the form of a grey or violet powder, which is the violet-coloured body occasioned by exposing chloride of silver to light. But this is not metal silver; it is only a combination of silver with chlorine, containing half as much chlorine as chloride of silver. Silver and chlorine form two combinations—one white and rich in chlorine; the other violet, and with little chlorine, named hypo-chloride of silver. In the same manner, silver forms two combinations with bromine—one light yellow, rich in bromine, named bromide of silver; and a yellowish-grey compound, less rich in bromine, named hypo-bromide of silver. Further, analogous to these there exist a yellow iodide of silver, and a green hypo-iodide of silver, less rich in iodide. Hypo-bromide and hypo-iodide of silver are produced exactly in the same manner as hypo-chloride of silver, through the operation of light ; therefore the chemist says that bromide, chloride, and iodide of silver are reduced to the corresponding hypo-salts.

The change of colour by which this chemical change is perceived is most striking in chloride of silver, less in bromide of silver, and least so in iodide of silver.

It would appear from this that chloride of silver is the most useful to photography. But the case is different. We have previously seen, in treating of the practical part of photography, that plates of iodide of silver and of chloride of silver are exposed in the camera. The image thus produced is virtually invisible, but becomes visible through a subsequent process, named the developing process.

In daguerreotypes the exposed plate of iodide of silver was fumigated in the vapour of mercury. In this case the vapour of mercury is precipitated in fine white globules on the exposed places, the amount varying according to the strength of the light. In the present treatment with collodion, the plate is washed over with a solution of green vitriol. This becomes mixed with the adhering solution of silver, and precipitates from it a fine black silver powder, which adheres to the exposed places of the plate.

Therefore, in both cases we have a finely pulverized body, which is attracted and retained by the exposed places—a mysterious process, as interesting as it is practically important.

From this it appears that it is by no means the colouring of the salts of silver which renders the image visible, but the subsequent developing process.

If an experiment be made simultaneously with chloride, bromide, and iodide of silver, by exposing and developing them, it is found that chloride of silver gives the feeblest picture under the developer, bromide of silver a stronger one, and iodide of silver the strongest. Therefore, the very body which was most strongly coloured by light is the least coloured under the developer, and the body which is the least coloured by the light, viz. iodide of silver, is the most coloured under the developer.

The developing process is of immense importance. If it were attempted to produce a picture by exposure in the camera, without *developing*, an exposure of hours would be required before the impression could be seen. The developing process permits, under favourable circumstances, the impression to become visible after an exposure of only one-hundredth of a second.

Pure iodide of silver was formerly used in photography, but iodide of silver is now used mixed with bromide of silver. This change was made because it was soon perceived that iodide of silver is very sensitive to *strong* light, but by no means so to weak light. For example, in taking a portrait, iodide of silver gives the light parts in a few seconds with great clearness, such as the shirt and the face; whereas the darker parts, such as the shadows, the dark coat, etc., are very feebly given. But, if some bromide of silver is mixed with iodide of silver, the coating of combined iodide and bromide of silver gives a weaker but still intense picture of the clear parts, while it gives a much better impression of the dark parts than iodide of silver alone.

The mixture of iodide and bromide of silver is effected

in practice by adding to the collodion a salt of iodine and a salt of bromine—for example, iodide of potassium and bromide of cadmium. Both are decomposed in the silver bath. Iodide of potassium and nitrate of silver produce iodide of silver and nitrate of potash, and in the same way bromide of cadmium and nitrate of silver produce bromide of silver and nitrate of cadmium.

A considerable quantity of the solution of silver remains also adhering mechanically to the collodion coating. This adhering solution of silver is by no means a matter of secondary importance; on the contrary, while washing in the developer, it affords the material from which the fine silver powder is precipitated that is necessary for the development.

If the developer (for example, a solution of green vitriol) is mixed with a solution of silver, the silver is precipitated in the form of a fine powder. For green vitriol is greatly attracted by oxygen, and, taking it up readily, passes into sulphate of iron. Accordingly, if a body containing oxygen (for example, nitrate of silver) is mixed with green vitriol, the latter withdraws at once the oxygen from the silver salt, and the silver becomes separated by the process. Other bodies that readily combine with oxygen operate in like manner, especially some substances from the organic world, such as pyrogallic acid, etc. It was formerly thought that green vitriol reduced the iodide of silver affected by light, and this erroneous opinion is actually found in some of the most recent works on chemistry. It can be easily proved that this view is false. For, if a plate is exposed and the nitrate of silver adhering to it is washed over, and then the developer poured upon it, no picture appears, which proves that green vitriol alone is not able to operate on iodide of silver exposed to light. But if a solution of silver is added, a picture appears immediately.

The solution of silver adhering to the plate plays again another part. If a plate is washed before it is exposed—that is, if all the nitrate of silver which adheres to it is removed, and it is then exposed—it will be remarked that it is far less sensitive than when the nitrate of silver is present. Whence does this proceed?

The matter is explained by the peculiar relation of many sensitive bodies.

There are bodies which in isolation are either not, or only very slightly, sensitive to light, but which become so when combined in a compound which is able to unite with one of the liberated constituents during exposure to light. For example, chloride of iron is not sensitive to light; but chloride of iron dissolved in ether is sensitive to light, because the liberated chlorine unites at once chemically with the ether.

The same remark applies to iodide of silver. This is, by itself alone, little sensitive to light; but if a body is present which can combine with iodide, it is quickly transformed in the light. A body of this nature is nitrate of silver, which absorbs iodine with the greatest ease.

This explains the greater sensitiveness of iodide of silver in the presence of nitrate of silver.

It follows from this fact, which was first accurately determined by the writer of this book, that other bodies which unite easily with iodine increase the sensitiveness of iodide of silver.

Among these bodies may be enumerated extract of copper, extract of tea, morphine, tannin; and consequently place in the hands of photographers the means to prepare what are called dry plates. The plates, which are prepared in a silver bath, only remain moist for a short time; the adhering solution of silver dries up, and then dissolves the iodide of silver, so that the plate is actually eaten into. Therefore it is not possible to keep a supply in a moist state, and to prepare them for any length of time, which would be very advantageous in travelling.

But dry plates retaining impressions are prepared by washing in water, and removing the nitrate of silver adhering to the moist plate, and then coating the plate with a solution of a substance having relation with iodine; for example, with tannin or morphine. Such coatings can dry up without injury to the film of iodide of silver, and in this manner a dry plate is obtained retaining durable impressions. I admit that the sensitiveness of these plates is considerably less than that of moist plates, but this is of no detriment in the case of objects emitting a strong light. The development of dry plates of this kind is commonly effected with pyrogallic acid. This is a substance that is obtained by dry distillation of the gall-nut. It operates very powerfully as a reducing medium; that is, it precipitates metallic silver from its solutions, exactly as green vitriol does.

But pyrogallic acid alone is not able to bring out an image on an exposed dry plate, because another substance is required for this purpose, yielding pulverized silver. This substance, viz. a solution of silver, is found on the plates themselves when they are moist.

But in the case of dry plates, the salt of silver has been washed off; therefore a mixture of pyrogallic acid and solution of silver must be employed as developer. Silver powder is precipitated from this, and, adhering to the exposed places, brings out the image into view. Nevertheless dry plates do not give such beautiful and secure results as moist plates.

We have therefore given an illustration of the photo-chemical phenomena in the production of a camera picture. The essential part of this process—the negative process—consists in the developing of an invisible light impression through a subsequent operation.

But all pictures are by no means prepared in this way. We have already seen, on the contrary, that the pictures on paper are occasioned by the production of a *visible* impression of light, a piece of sensitized paper being exposed until it is coloured dark. In this case no developing is required. The picture is exposed to the light till it has received the necessary consistency.

The process put in practice in this case is quite simple. The positive paper contains chloride of silver and nitrate of silver. The former is quickly, the latter slowly, reduced by the light; that is, precipitated as metallic silver, which is separated as a brownish powder. Chloride of silver would only be reduced to hypo-chloride of silver. But by the presence of paper-fibre the process of reduction is carried further; it produces metal silver. Then the chlorine set free by the light combines immediately with the silver of the nitrate of silver, and produces chloride of silver again. This is immediately decomposed by the light. A fresh quantity of brown metallic silver is thus separated, here again free from chlorine; and this process is repeated as long as nitrate of silver is present, and as long as the light operates.

Pure chloride of silver alone only offers a faint impression, but in contrast with nitrate of silver it presents a very vivid image. The picture, in the form in which it is produced by the light, is not durable—it would turn brown through the further operation of light on the white places; and to prevent this, the salts of silver still adhering to the paper, sensitive to light, must be removed. The nitrate of silver is removed easily by washing with water, for it is soluble in water; but the chloride of silver must be removed by plunging in a solution of hypo-sulphite of soda. This salt becomes transformed with chloride of silver, forming chloride of sodium and sulphate of silver, and the latter combines with the hypo-sulphite of soda to form a sulphate of tartar; and this sulphate, remarkable for its peculiar sweet taste, is soluble in water, and can be removed by washing.

If a fresh impression is plunged in hypo-sulphite of soda, it suddenly changes its beautiful violet colour—it becomes of a yellowish brown, and this tint is not liked. It does not interfere with the effect in technical and scientific pictures, but is a great drawback in portraits and landscapes; so that the positive prints are subjected to a further treatment, styled the colouring process. To this end it is plunged in a very diluted solution of gold. This solute contains chloride of gold. Metal silver has more affinity with chlorine than gold; hence it combines with the chlorine, forming chloride of silver, while the gold is precipitated. It becomes separated in the shape of a blue colour adhering to the outlines of the picture, and this blue, mixed with the brown of the picture, gives a pleasant tone, which does not change in the fixing-bath,—that is, in hypo-sulphite of soda.

Accordingly, every paper photograph consists of silver and gold, in the proportion of four parts of silver to one of gold; the quantity of both substances being very small. In a picture of 44×47 centimeters, or 17 inches by 22, only one-thirteenth of a gramme, or 1·187 grains, of metal silver are contained. Its value is about one German pfennig,* and the value of the silver in a carte de visite is about one-thirtieth of a farthing. The question may here arise, how it happens that photographers charge so high for their pictures? A sufficient reply is found in the fact that the price is not determined by the value of the materials, but by the labour which has been necessary to produce the pictures. It must be remembered that a photographer has to make twenty-eight operations to produce a negative, and eight to produce a positive; that a picture is often a failure; that four-pennyworth of salts of silver must be employed, besides one farthing's-worth of silver in preparing a sheet, and that at the utmost only one-third of this silver can be recovered from the bath. Nor should it be forgotten that the paper itself is valued at threepence, that cardboard of the same value is required for the mounting, and that further outlay is needed for hire of premises and assistants—all which circumstances certainly justify the price demanded.

If it is borne in mind that thirty-three times as much silver must be employed as that which actually remains when the picture is finished, it will be seen that the amount of silver consumed annually in photography must be enormous. It is valued at about £350,000.

* About the value of an English farthing.

CHAPTER XII.

ON THE CORRECTNESS OF PHOTOGRAPHS.

Influence of the Individuality of the Photographer—Different Branches of Photography—Influence of Lenses, of the Length of Exposure, of Colours and Models—The Characteristic Feature in the Picture —Deviation from Truth in Photography—Difference between Photography and Art.

(a) *Influence of the Individuality of the Photographer.*

In the previous chapters we have become acquainted with the development and the theory and practice of photography. We have mentioned cursorily various practical applications; for example, the *licht-paus* process. It is our present purpose to give our special attention to one point which is of great import in judging of the value of a photograph.

Most persons have a fancy that the application of photography is always uniform, whatever may be the object to be taken, and, therefore, that a photographer who can take a portrait must be able to take equally well a machine, a landscape, or an oil-painting. This results from the erroneous notion that the picture makes itself when the photographer opens and shuts the lid. But our readers know already that the picture does not make itself, but that it must be first developed, brought out, fixed, and copied. In all these operations there is no precise measure or rule how long the photographer should expose to the light, develop, fortify, copy, and tone the picture. This depends on his option and judgment; and he is able at pleasure to bring out the picture more or less in detail, according to the time of exposure. Again, he can make it more or less brilliant, according to the degree of strengthening; he can make it more or less dark, according to the mode of im-printing; more or less blue, according as he tones it down. But what is it that directs his judgment to determine if the picture is correct or not? It is nature, and nature alone! He must know nature, and compare it with his picture. Nor is this easy. Nature appears positive to him, but in the picture she appears first negative; and if he compares the two, he must be able in his mind to convert the picture, that is, to change it and represent it as a positive, which it is afterwards to become. More comparing and study are required to do this than is generally supposed.

If two printed proofs are presented to a man who is ignorant of the art of printing, one of the sheets in question being well and the other ill printed, if the defects be not too glaring, this person will not be able to detect any difference between the proofs. Far other-wise is it with the practised eye of the printer, who immediately detects that in one proof the type is too thick, or thin, or leaded, or that the letters are faint, or blotched, or uneven. In like manner, a practised eye is needed to judge a photograph—an eye not only able to detect the finest details of the picture, but also the peculiarities of the original. The unprofessional man often uses the expression, "I have no eye for it,"—that is, "I am not accustomed to see such things,"—and it is in this manner that we first discover how imperfectly we use this, the most perfect of our senses.

A man born blind, and who recovers his sight by an operation, cannot at first distinguish a cube from a ball, or a cat from a dog. He is not accustomed to see such things, and must first exercise his eyes and learn to see.

We, also, though in possession of sound organs, are blind to all things that we are not accustomed to see; and this fact is most apparent in art, as also in photography, so closely related to it.

If photographers principally engaged in taking portraits are not able to produce a good landscape, the reason of this is that they have no eye for landscape —that they consider a picture to be good after too short an exposure, or when imperfectly developed and strengthened, or when inaccurately printed. It proceeds from their not knowing the influence exercised by the position and intensity of the sun to the aerial perspective produced by clouds, without speaking of other points of less importance.

Thus every class of subjects requires a special study, though the manipulation of photography remains in all cases the same; therefore, there are photographers whose proper province is portraits, and others devoted to landscapes, to the reproduction of oil-paintings, etc.

(*b*) *Influence of the Object, of the Apparatus, and of the Process.*

The remark is frequently made by admirers of photography, that this newly-invented art gives a perfectly truthful representation of objects, understanding by the term truthful a perfect agreement with reality. Photography can, in fact, when properly applied, produce truer pictures than all other arts; but it is not absolutely true. And, as it is not so, it is important to become acquainted with the sources of inaccuracy in photography. Many exist. I shall treat here especially of optical errors.

The lenses which are employed in photography do not always give absolutely true pictures. Suppose, for

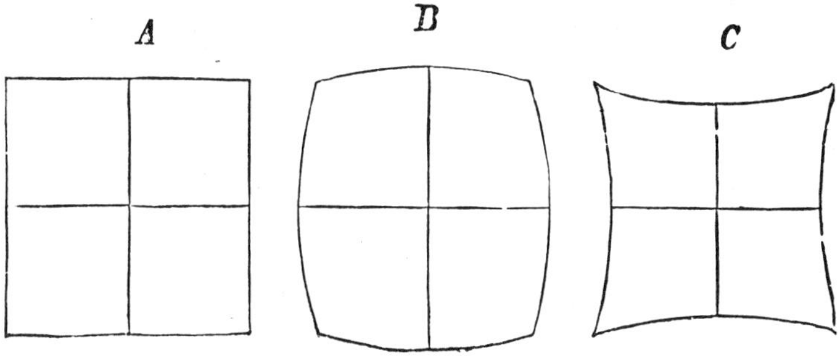

Fig. 50.

example, that a simple lens receives the impression of a square; it often represents it with curvilinear sides, as in the accompanying diagrams, though with a feebler outline. A picture thrown off quite out of drawing by such a lens, in which straight lines turn out as curves, is evidently inaccurate. The inaccuracy may not be felt by many, but it exists. It may perhaps be expected that this defect disappears in the case of what are called correct lenses, but let the attempt be made to obtain a view with these correct lenses of lofty buildings taken from a low position. The lines that ought to be perpendicular commonly converge upwards. This is caused by the photographer being obliged to direct his instrument at an acute angle upwards, in order to be able to take in a view of the whole building. In doing this, perpendicular lines project themselves, converging upwards. To avoid this defect, lenses have been made with a very large field of view. These are called pantoscopes. But these reproduce distant objects apparently

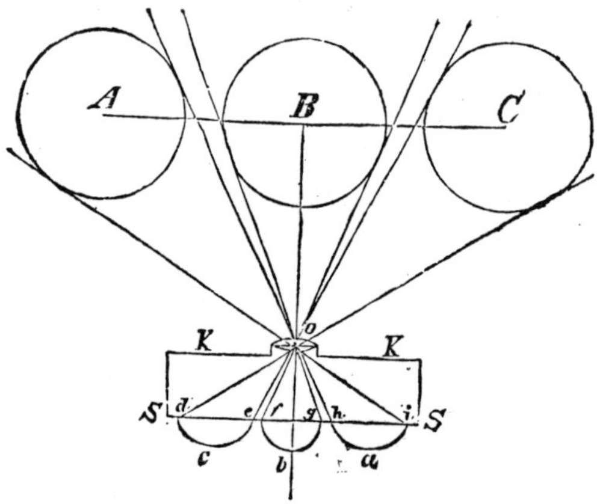

Fig. 51.

on a very small scale, and objects near at hand on a very large scale,—peculiarities unnoticed by unprofessional persons, but detected by close observers of nature.

A remarkable phenomenon, exciting the wonder of the uninitiated, is the distortion of spheres in photography. Let the reader imagine a row of cannon balls; these will always appear balls to us, and the artist will always draw them as a circle. But if they are taken through a lens with a large field of view, the balls situated near the rim of the lens no longer appear circular, but elliptical.

To explain this phenomenon,.we must attend once more to the mode in which the picture is produced. Let it be conceived that there are three balls *A B C* in front of a camera *K*, with the lens *o* (Fig. 51). Each ball projects a cone of rays on the optical centre of the lens. This is continued within the camera, and cuts the surface of the picture, if its axis falls obliquely upon it, in the form of an ellipse, such as *A C*. Only, if the axis of the cone of rays is perpendicular to the surface of the picture *S S*, as at *B*, the picture appears a circle. I admit that this defect only occurs when the field of view of the lens is very large, and the balls are situated very near its rim.

A photographer brought to the author the picture of a castle having a row of statues in front of it, which he had taken with a lens having a large field of view. Singularly the heads of the statues towards the margin became continually broader, and similarly their bodies; and the slim Apollo of Belvedere, who unfortunately stood on the very edge of the margin, had such full-blown cheeks and so protuberant a paunch, that he looked like Dr. Luther.

But, quite independently of these considerations, there is another point that must materially affect the accuracy of photographic representation. Photography generally gives the light parts too light, and exaggerates the dark shadows. This is a fundamental error which is associated with their very nature, and which it is very difficult to avoid. It is seen in the most evident manner in taking objects lighted by a brilliant sun; for example, a statue. If the exposure is short, a detailed picture is obtained of the light side, but the shady side is a black daub or blotch. If the exposure is long, the shady side is full of detail, but the light side exposed too much, and so thickly covered that the details are wanting in it. Hence photographers are often driven to subterfuges if they wish to obtain a correct picture; they are obliged to mitigate the contrasts—to make the light more toned down, and the shades lighter than painters are wont to make them. The latter often exclaim when they see the photographic exposure of a model, and wonder if the picture will be correct. And no doubt, in the case of landscapes and architecture, the results are not always satisfactory.

The author once took a photograph of the interior of a laboratory. It presented the appearance of an ordinary vaulted hall. All was quite excellent. The tables, stones, retorts, lamps, etc., were all seen, only the vaulted ceiling was quite dark. New attempts were made, with exposures of 20, 30, or 40 minutes. At length a trace of the vault appeared; but now the objects in the vicinity of the window were suffering from too much exposure; that is, they had become as white as if they had been snowed over. This circumstance of photography exaggerating the dark parts appears again in very simple matters, such as the reproduction of copper-plates. A photographer once reproduced a painting of Kaullach's "Battle of the Huns." He produced a charming photograph, but the city in the background appeared too thick and black, and not sufficiently toned off. The customer refused the photograph and demanded another. The photographer made another attempt, giving a longer exposure, and now the distance appeared softened down; but, unfortunately, the objects close at hand, which had to appear black and clear, turned out grey. In the end, the photographer escaped from the difficulty by negative *retouche*. These are quite ordinary examples to show how difficult it is to reproduce an object correctly.

But we come now to the worst point, that of colour. Photography gives the cold colours—blue, violet, and green—too light, and the warm colours too dark. Take as an instance the photographs on sale of "Sunset on the Ganges," by Hildebrandt. It represents a red glowing sun with clouds of chrome yellow on an ultramarine sky. But what becomes of all this in the photograph? A black round disk between black thunderclouds. It looks like an eclipse at Aden. The difficulty of representing nature is still more patent when the photographer attempts to grapple with higher artistic questions. Let us take an example. There exists a pretty *genre* picture called "A Mother's Love." A mother sits reading in an armchair; her little darling embraces her suddenly from behind, and, delightfully surprised, she drops her hand with the book, turns to look at her little pet, and offers her cheek to the little boy to kiss.

A photographer was inspired with the idea of producing a similar picture with the help of a living model. He found a comely maiden, who agreed to personate the mother, and a good-looking boy was also found. An armchair for the mother, a chair, and other suitable furniture were easily procured. The next point was the grouping. The pseudo-mother was very accommodating to the requirements of the photographer, and even assumed a look which, for want of a better, might pass as the expression of a mother's love. But the boy was not of the same mind. He was by no means attracted by the pseudo-mother—he protested against coming near her, and a good cuff was needed to make him take up the requisite position. Time was thus lost. The mother began to feel uncomfortable in the irksome position, straining her neck. The photograph was taken at last, and turned out sharp and without spot or blemish. The models were dismissed to their great satisfaction. What was the result? The boy was embracing his mother with a face bearing evidence of the cuff he had received, and with a look as if he would have liked to murder her; and she regards him with an expression that seems to say, "Charles, you are very unmannerly," and appears greatly annoyed that her pleasant reading has been interrupted. Can it be said that a picture of this kind correctly expresses the intention of the artist? Does the picture thus produced correspond accurately to the legend, "A Mother's Love"? The untruthfulness of such a picture will be evident to every one.

Thousands of pictures of this class are offered for sale. About ten years ago errors of this kind were committed by the thousand in stereoscopic views, and if they meet with approval this must be referred exclusively to the bad taste of the public. But it may be said in this case it is not the photographer who is guilty, but the unwilling models.

Nevertheless, it is this very circumstance that throws such immense difficulties in the way of taking good photographic portraits. Many persons by no means wish that their characters should be correctly given.

The rascal wishes to appear an honourable man in his picture; tottering old men desire to appear young, foppish, and lively; the maid-servant plays the fine lady in the atelier; the tradesman's daughter would be a court lady, the street-sweeper a gentleman. Thus the picture serves them only as a means of flattering their personal vanity; and, in order that these people may appear very noble and distinguished, they put on a Sunday's dress, often borrowed and a very bad fit. They practise at home, moreover, before their looking-glass, in the presence of papa, mamma, wife, or lover, impossible attitudes in an artistic point of view. Even cultivated persons are not exempt from these absurdities. Thorwaldsen relates of Byron, who gave him a *séance*, "He sat down opposite to me, but assumed, immediately I commenced, a perfectly different expression. I called his attention to this. 'That is the true expression of my face,' replied Byron. 'Indeed,' I rejoined, and then made his portrait exactly as I wished. All persons declared my bust to be an excellent likeness. But Lord Byron exclaimed, 'The bust does not resemble me; I look much more unhappy.' The fact was that at that time he wished to look intensely miserable," adds Thorwaldsen. The photographer is even in a worse case. If Byron had come to a photographer and had presented his face of misery to the camera, what could the photographer have done? He is unfortunately dependent on the model, and many models leave him in the lurch at the critical moment, often not intentionally, but from nervousness or inadvertence. Much depends here on the influence of the photographer, who must know how to control his sitters with courtesy;

but many portraits fail without any fault on his part. The author has often witnessed how persons of his acquaintance, at the moment of being taken, assume quite a strange expression without being in the least aware of it.

There are still more characteristic cases of photographic inaccuracy which cannot be attributed to the models. Let us suppose that a photographer, stimulated by the beautiful pictures of Claude, Schirmer, and Hildebrandt, wished to photograph a sunset. He evidently can only expose his plate for a moment to the dazzling bright sun. What sort of picture is the result? A round white blotch and some shining clouds around it. That is all that appears clearly. All objects in the landscape—trees, houses and men—have had too short an exposure, and form a black mass. There, where the eye clearly distinguishes road, village, forest, and meadow, it sees in the photograph nothing but a dark patch without any outline. Is such a picture true? Even the most fanatical enthusiast of photography will not dare maintain this.

Such cases, where violent contrasts of light and shade make the production of a correct picture quite impossible, are countless in number. Let any one examine the majority of the photographs of the white Royal Monument in the Thiergarten at Berlin. The monument is excellently given, but the background of trees is a confused black mass, without details, without shades of tone; the architecture and other features are there, all except the splendid foliage that delights the eye at that spot. Still more numerous are the photographs of rooms, in which the dark corners, quite discernible to the eye, present nothing but pitchy black night. There are other cases besides these of photographic incorrectness.

Suppose we are looking at a mountain landscape. A small village, enclosed on both sides by woody hills, occupies the centre, its houses extending along the declivities and scattered picturesquely among the trees. A ridge of finely-broken mountains in the background, their summits shining in the setting sun, frame in the wonderful picture, whose effect is only injured by one object—a ruinous pigsty close to the spectator, with a dungheap beside it. A painter, wishing to paint this scene, would certainly have no scruple about altogether leaving out the pigsty, or leaving it so indistinct and dark that it would not injure the landscape. But what is the photographer to do? He cannot pull down the offending object. He seeks another position; but there the greater part of the landscape is concealed by trees. He ends by admitting the pigsty, and what kind of picture is the result? On account of its vicinity, the pigsty appears of colossal size in the picture. On the other hand, the landscape, which is the principal thing, appears small and inconsiderable. A still more fatal adjunct is found in the dung-heap occupying almost one-fourth of the picture. As the most brightly lighted part of the photograph, it immediately attracts the eye of the beholder; it diverts his glance from other important points; it acts as a disturbing influence. The photograph obtained does not appear as a picture of the landscape, as it ought to be, but as a view of the pigsty. The accessory has become the principal point. The picture is untrue. It is untrue, not because the objects it represents were not present in nature, but because the accessories are presented too glaringly and too large, while the principal parts appear too small, indistinct, and inconsiderable.

This brings us to a weak point in photography, which represents accessories and principal features as equally defined. The plate is indifferent to everything, while the genuine artist, in reproducing a view of nature, gives prominence to what is characteristic, and entirely keeps under or softens off accessories. He can dispose and manage it with artistic freedom, and he has a perfect right to do so, because, by his giving prominence to what is characteristic, and dropping what is accessory, he is truer than photography, which gives equal prominence to both, and often more to what is accessory. Reynolds says of the portrait of a lady in which an apple-tree was most carefully painted on the background: "That is the picture of an apple-tree and not of a lady." Similar remarks might be made on seeing many photographs. It is a cardinal error in their case, that they give a stronger tone to accessories than to essentials. They present a conglomerate of furniture, and it is only after careful inspection that a man is detected sticking among it, whose portrait is to form the picture. In another case a quilted white blouse is seen, and it is only after some time that a girl's head is perceived rising above it. A park is seen in a landscape, with fountains and other adornments, and it is only after some time that a black coat is seen confounded with an equally dark bush.

It may perhaps excite surprise that the writer ascribes greater truth to painting than to photography, which is generally regarded as the truest of all methods of producing pictures. It must be self-evident that the remark has only been made in connection with works of the first masters. One of the great services of photography is that it has rendered impossible those daubs of inferior artists formerly offered for sale in every street. But the perfect picture of the photographer is not self-created. He must test, weigh, consider, and remove the difficulties which oppose the production of a true picture. If his picture is to be true, he must take care that the characteristic is made prominent and the accessories subordinate. The non-sensitive plate of iodide of silver cannot do this. It receives the impression of all that it has before it, according to unchangeable laws. The photographer attains this end, on the one hand, by appropriate grouping of the original; on the other, by a proper treatment of the negative. I admit that to do this; he must also be able to detect what is characteristic and what accessory in his original.

Therefore, whoever wishes to undertake any photographic production must first become familiar with the object that he wishes to take, that he may know what he has to do. The photographer will not, indeed, be able to control his matter, like the painter, for the disinclination of models and the optico-chemical difficulties often frustrate his best endeavours, and hence there must always be a difference between photography and a work of art. This difference may be briefly summed up by saying that photography gives a more faithful picture of the form, and art a more faithful picture of the character.

CHAPTER XIII.

LIGHT, SHADE, AND PERSPECTIVE.

The Difference between the Picture and Reality—Effect of Shade—Perspective Foreshortenings—Effect of the Position of the Spectator—Influence of Distance—Influence of the Eye-line.

In the previous chapter, while treating of the incorrectness of photographs, we tacitly took for granted that it is possible to give a true picture of an object, if not by photography, by the hand of a skilful artist.

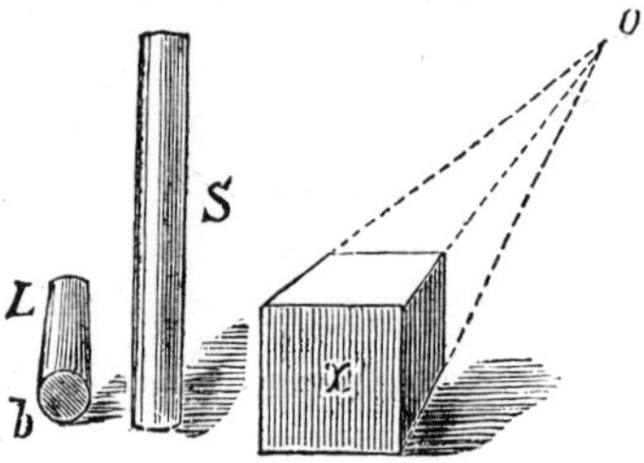

Fig. 52.

We will now see how far this assumption is admissible.

Let the simplest case be taken; for example, a cube or a cylinder. Let these be drawn, and figures will be obtained nearly identical with those marked *X* and *S* in the diagram. Now, these figures are flat like the paper, while the originals are solids. It may be said that picture and solid agree; but it is not so. Let a blind man be questioned, who knows the bodies by touch. Now, the cube can be moulded in marble or gypsum.

In this case, the deception—for such it is—can be carried to great lengths. The wood of the cube or of the cylinder can be imitated by painting. The eye will readily pronounce such imitation to be wood. The blind man, who feels both, will say: The form agrees, but not the mass—one cube, that of wood, feels warm; the other, that of stone, cold.

The principles that apply to these two objects apply to all objects and their representations. None of them is a perfectly true copy of the object. When the surface representation makes on our eye the impression of a solid object, this is a deception, by which our eye suffers itself to be deceived.

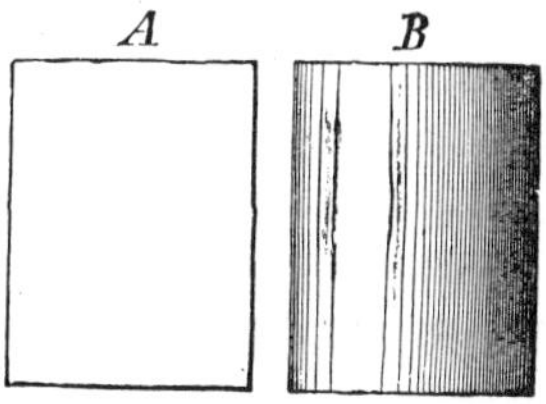

Fig. 53.

If two rectangles are drawn, *A* and *B*, on paper, both appear as plane figures. But directly the rectangle *B* is shaded with thinner or thicker lines, the rectangle no longer appears such, but a spherical body. Thus, by imitating the gradations of light and shade, we have produced a deception for our eye, and this aid—the division of light and shade—is one of the most important distinctions in art, to give a spherical appearance to flat surfaces.

But there is another and a more important means of deception—perspective.

If we consider a cube (Fig. 52) whose faces are of equal length, we perceive that these faces appear of very different length. The surface turned towards our eye appears a square, while the others are shortened in a marked degree, the surface appearing quite irregular, the parallel lines running into one, and converging to one point *o*, called the *vanishing point* (Fig. 52). The same thing happens with all other bodies: a pendant human arm or a standing column *S* (Fig. 52) appear at their full length, but the lying column *L* and the arm extended towards us appear *foreshortened.* Their dimensions are contracted; in short, we see, instead of the shaft of a column, only its circular base *b*, and this again appears sometimes round, when its full surface is turned towards us, at others an ellipse, which it is not in fact, and in this case the parallel sides of the column run into one. The track of a railroad viewed in perspective presents the same features. The fact that we do not feel this deception—for such it is—to be one, results from habit.

We know from experience that the arm extended towards us is longer than it appears in perspective, to our eye, and also that the rails which appear to run together are parallel. We are continually correcting the errors of our visual organ. Accordingly, the eye gives us a false representation of objects, and the painter takes advantage of this circumstance. He represents the lying column *L b*, and the retiring sides of the cube, as falsely as we see them—that is, "foreshortened" in their dimensions, running together in their parallel lines —and every one is deceived by this.

It is the province of the artist as of the photographer to represent *perspective* correctly, that is, as it appears to our eye. If this is not the case, the picture appears incorrect.

Perspective teaches us the law of foreshortening.

Our eye is a camera obscura with a simple landscape

lens. It is known, from optics, that the representation of a point lies on the straight line drawn from the point to the optical centre of the objective. The representation of the point is at the place where this line, named the principal radius, cuts the plane of the image—the ground glass shade of the camera, or the retina of our eye. Accordingly, the representation of a straight line is the place where the radii from the separate points of the line, passing through the optical centre, cut the ground glass table or shade. Now, these radii form a *plane*, and this plane cuts the *flat* table of the picture in a straight line. Therefore the picture of a

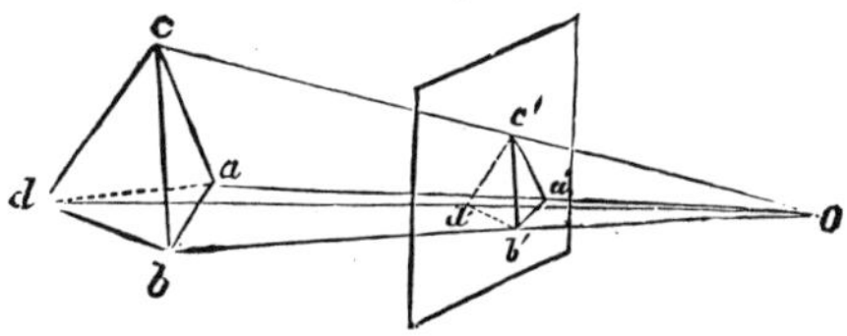

Fig. 54.

straight line is to our eyes another straight line, and the image of a plane triangle another plane triangle. If the flat figure is parallel to the retina, that is, to the table of the picture, by well-known stereometric laws the representation is like the original. Let the reader imagine a glass slab placed perpendicular to the axis of his eye; then the rays or pencils of light *a b c d* issuing from this object will cut it so as to form a figure *a' b' c' d'* (Fig. 54). If such a figure is constructed for a given point of intersection and a given canvas, this drawing, if brought to a proper position and distance from the eye, will produce on it exactly the same impression as the object itself. This is the secret of the deception that a plane picture, properly constructed, can appear spherical. A picture designed in the manner just described is named a *drawing in perspective.* It is evident that such a drawing must be viewed under the same conditions as those in which it was designed.

If *A B C D* (Fig. 55) is the outline of a house, *B* the canvas, *O* the point of intersection of the rays, *a b c d* the representation of the points *A B C D*, the eye must be brought exactly to the point of intersection *O* if the representation in perspective *a b c d* is to produce the same impression as the object.

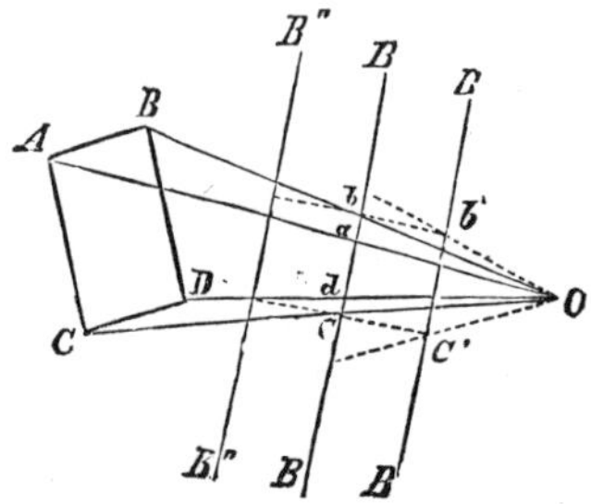

Fig. 55.

If the canvas is brought nearer to the eye (for example, to *B*), it is evident that the rays will intersect at a very different angle from those issuing from the object *A B C D*; accordingly, they cannot produce a correct impression. The same thing would be the case if the canvas were removed farther from the eye (*e.g.* to *B″*). Therefore every drawing in perspective must be viewed from the point of intersection of the rays adopted as the basis of its construction, if it is to produce a correct impression.

Now, photography is a drawing in perspective whose point of sight is in the objective. Accordingly the inspecting eye must be brought to the same distance as the objective, that is, to the focal distance. If this is not done, the impression is untrue.

We have lenses with a focal distance of only four inches, and even less; and at such a distance it is impossible to see a drawing with the unaided eye. To do this, it must be held at the distance of at least eight inches, and that is the reason why photography in such cases produces an untrue impression. Such cases frequently occur when views are taken with divergent lenses.

There are other abnormal appearances which accompany portrait taking. Thus, the same object presents an entirely different picture according as it is viewed far or near. Let the reader conceive a pillar with the outline *A B C D*, let it be viewed from *P*; in this case the faces *A B* and *C D* will be perfectly seen. Now let the spectator approach nearer to the object, for example, to *O*. From this position nothing is any longer seen of the faces ; the entire character of the picture becomes changed. If instead of a pillar a human face

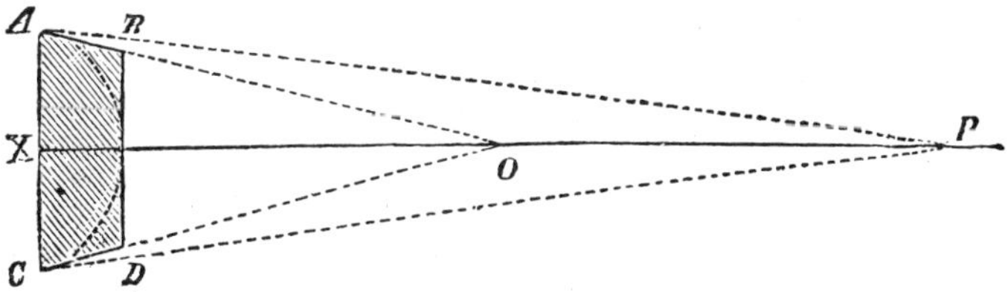

Fig. 56.

be thought of, it is evident that the cheeks will contract if we approach the object, and the face appear too narrow in proportion to its height.

The accuracy of this conclusion is proved by the following illustration. These are two representations of the head of Apollo at the distance of 47 and 112 inches.* The bust was placed perfectly upright, also

* In order to secure a correct reproduction of both, they were reproduced upon wood by a process of photographic wood-cutting. I admit that the reproduction does not give the powerful impression of the original, but it is sufficiently intelligible to the careful observer.

the photographic apparatus, and the directing line was most carefully measured.

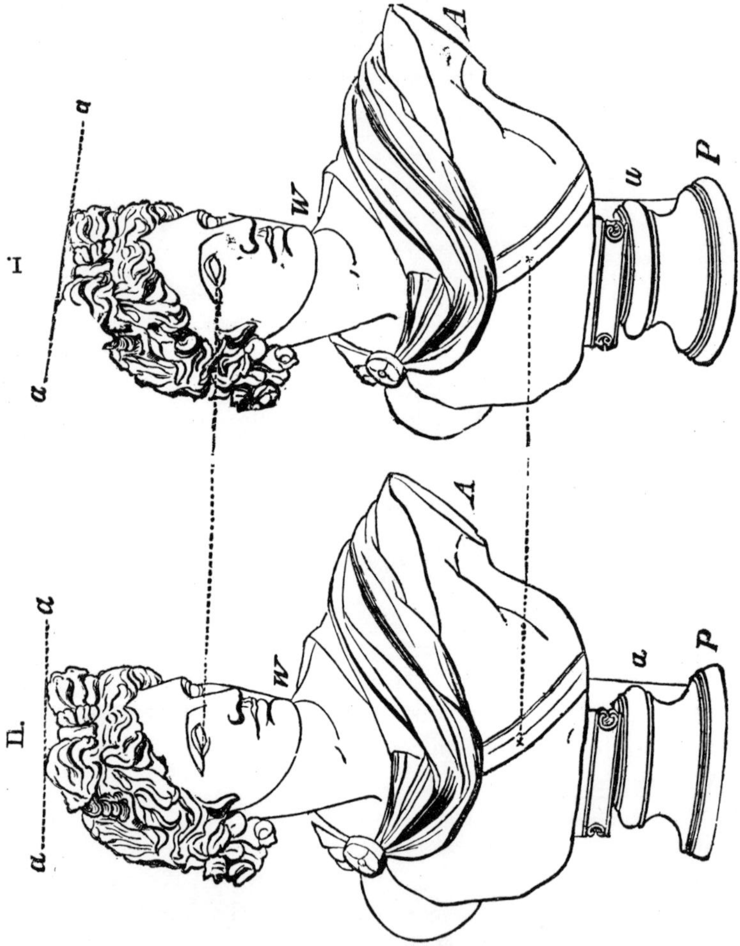

The contrast is obvious. The whole figure appears in I. slimmer, the chest almost contracted; on the other hand, the same model II. appears with full cheeks and square shoulders. That this slimness is by no means a mere deception of the eye is certain from the best measurement.*

The distances between the eye and the point on the chest marked by a cross are exactly equal in both busts —the greatest breadth of chest, including the ends of both arms, amounts in I. to 2·29472 inches, and in II. to 2·32283 inches. Quite independently of this glaring difference, there are other marked distinctions between the two heads which strike the careful observer. Let a line *a a* be drawn across the top of the hair—in II. it is horizontal, in I. it inclines to the right.

Next let attention be directed to the pedestal. The curves in I. are strongly inclined, and in II. are quite horizontal.

Let the ends of the arms *A A* be next considered. In I. the side surface is scarcely seen, and in II. it is very apparent. In like manner, it is clearly seen that the back pediment at *u* in II. stands out more than in I. In II. the head stands more between the shoulders (let the angle of the neck be observed at *W*), in I. it rises up more; therefore the whole form appears in I. to raise up the head more on high. In II. the head appears somewhat bent forward; and yet the figure was immovable, the lenses employed free from flaw, the eye-line and height were the same in both. Nothing was different but the distance.

The author, besides taking these two heads, has taken two others at the distance of 60 and 80 inches; and if

* In the original photograph, where the two busts stand out from a black background, this difference is still more marked.

the four heads thus taken are placed beside each other, it is seen how with the increase of distance the form becomes thickset, fuller, and dumpy; how the hair sinks more and more; how the ellipses become closer; and how the chest increases in width, and the stumps of the arms widen out.

Thus, therefore, we see very different views of the same object at different distances; just as the same portrait placed in different lights expresses an entirely different character.

It may be objected that these are small matters, and that it is indifferent whether it looks a little too thin or too fat. To many this may appear unimportant in the case of Apollo—most persons do not know in the least how he looks. But it is a different matter in the photography of portraits, where the highly distinguished personality of the customer is in question. Persons quite untutored in art have a very quick eye where their own physiognomy is in question—a line, a wrinkle, an outline, a curl, are in this case criticised, and differences that would not be at all remarked in the images of Apollo become very striking. It is therefore the affair of the photographer to attend to the effects of distance.

Now, many persons would perhaps wish to know, which distance is the best? which gives the most correct picture?

We might reply that this depends on the individuality of the person. Painters in general recommend for the drawing of an object a distance that is twice its own length; accordingly, for a man five German feet in height, a distance of about ten feet; for his bust, about five feet. The painter, however, has here greater freedom he can add, omit, and change at his pleasure. In photography this is only partially possible.

Just as elevated solid bodies appear different at different distances, hollow shapes also appear different at different distances.

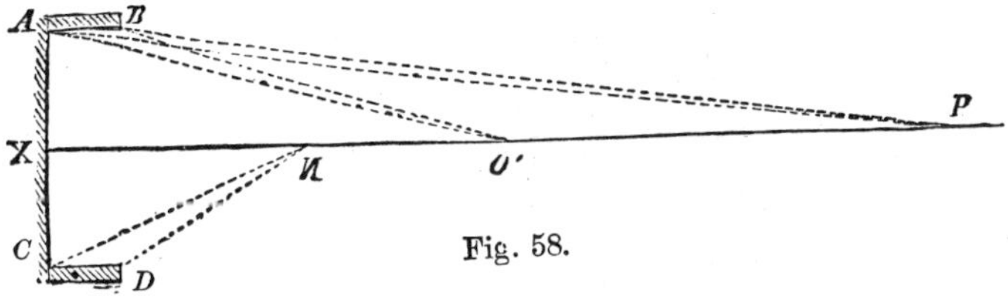

Fig. 58.

If *A B C D* (Fig. 58) is the inside of a box, we see the side *A B* from *P* much more foreshortened than

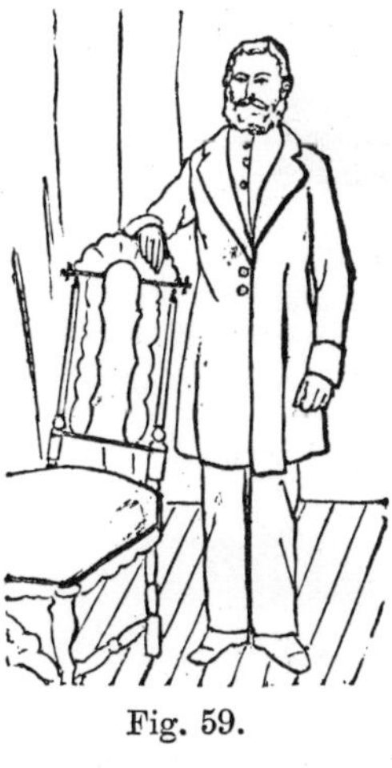

Fig. 59.

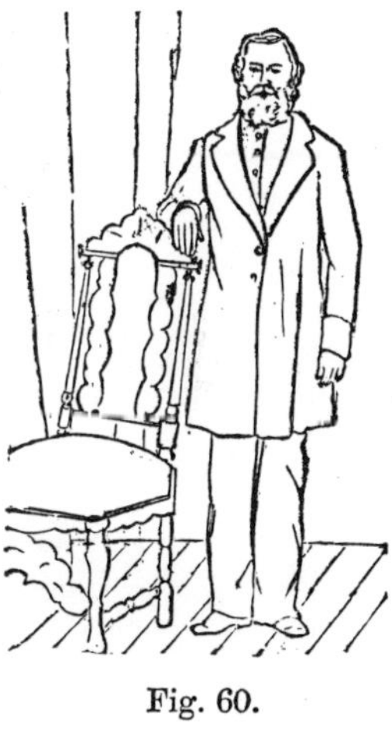

Fig. 60.

from *O'* to *N*; therefore, if taken under similar relations, from near and at a distance, it will appear wider in proportion to its height in the former case.

This state of things occurs if we imagine *A C* to be the trunk, and *C D* the lap or the feet, of a seated person. In that case the lap appears much larger in relation to the trunk, and the feet of a standing person facing you

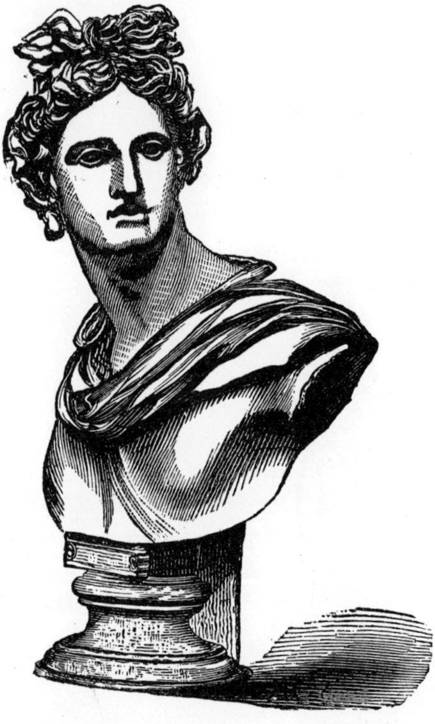

Fig. 61.

appear longer from *N*. Let the reader observe, for instance, the foot of the Apollo in I. (Fig. 57), which is much more prominent than in II. Lastly, let *C D* (Fig. 58), be supposed to be the carpet or ground; this will appear wider—that is, rising higher—seen from *N*. Therefore, if the same person is taken from different positions, *P* and *O′*, so that the height of the body remains the same in both pictures, in that one taken at a shorter distance the prominent parts—lap, hands, and

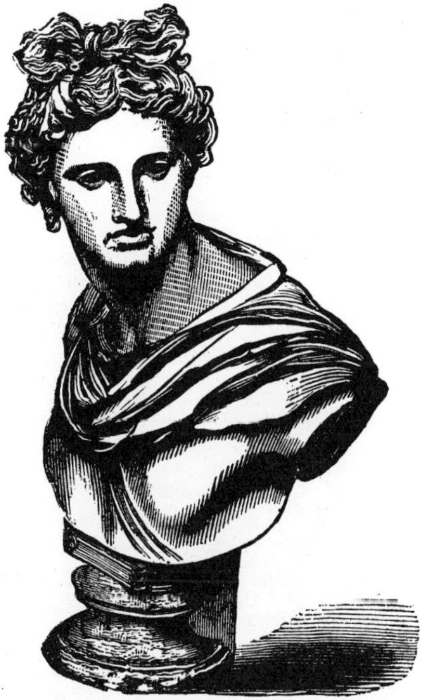

Fig. 62.

feet—appear wider, and the ground or chair more inclined (Fig. 59) than in the picture taken from *P*.

Very essential changes result from an alteration in the height of the spectator's eye.

If a standing person is looked up to, so that the head

of the spectator is lower than the head of the object, the latter appears thrown back. If the head of the spectator is on a level with the head of the object, the latter appears perpendicular; if the spectator is higher, the head of the object appears inclined forward.

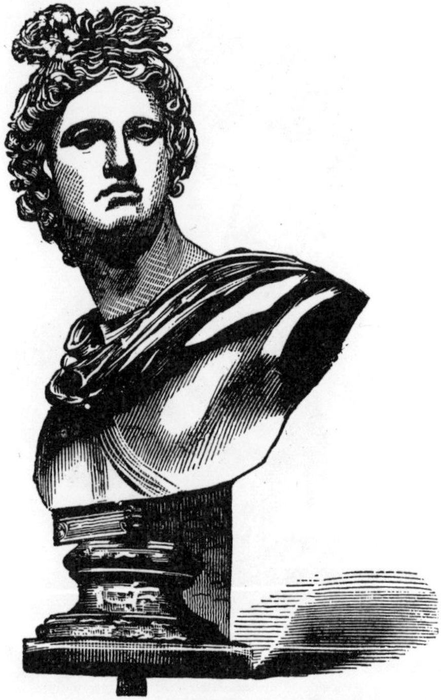

Fig. 63.

The three accompanying diagrams, taken from photographs, will make this evident. The first shows the view taken from a level, the second the view taken from above, the third the view from below.

Similar differences occur in viewing a landscape from

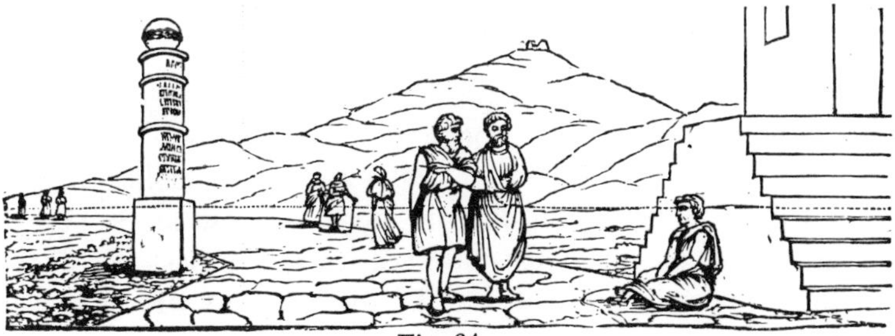

Fig. 64.

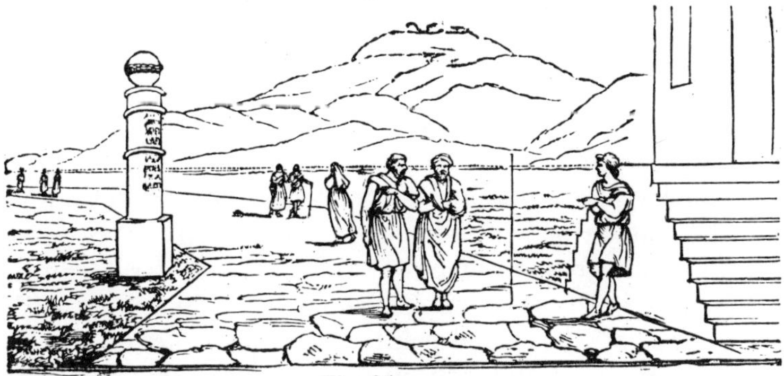

Fig. 65.

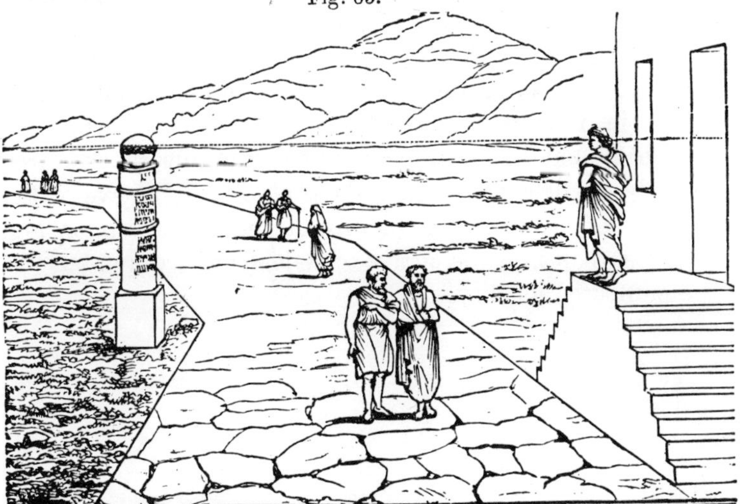

Fig. 66.

a high and from a low position, as may be seen from the three accompanying woodcuts. The dotted horizontal line shows the height of the eyes of the looker-on (his horizon). The first picture gives a view as a person sitting on the ground would see it: the milestone on the left appears unusually high, towering to the sky, and the men appear taller, but the ground looks contracted (foreshortened). The second picture gives the view as seen by a man standing erect: in this case the ground widens out, rising higher, and the milestone appears lower. In the third picture, which gives the view from twice the height of the man, the figures and the milestone appear small and contracted. You look down upon them as on persons who are smaller than the spectator, while the ground widens out and rises considerably in the picture. These examples show how important is the choice of the stand-point both in photography and painting, and how an incorrect choice of it produces quite an abnormal view of the objects. The photographer is unfortunately often obliged to take a position that does not give a favourable view, for example, of lofty buildings in narrow streets (the Rathhaus in Berlin, St. Stephen's at Vienna), or among mountains, where often the trunk of a tree, which did not offend the eye of the spectator, destroys the picture of the photographer, and forces upon him the choice of a less picturesque but unincumbered locality.

CHAPTER XIV.

THE APPLICATIONS OF PHOTOGRAPHY.

We have by this time learnt to know the difficulties which oppose the obtaining a correct photographic picture; and now the reader will more easily understand, if we examine with greater minuteness the different problems the solution of which photography has up to the present time attempted.

We shall prolong this examination only so far as it is useful for the comprehension of the subject, and as it is of general interest to every one.

Section I.—Portrait Photography.

Popularity of the Portrait Branch—Æsthetical Defects—Dependence of Success on the Person to be taken—Effect of Dress—Effect of Colours—Pictures of Children and in Groups—Effect of the Size of the Picture—Life-size Pictures—Momentary Pictures—Photographic Copies of Photographs.

Scarcely any other branch of photography has enjoyed so much popularity as that of portrait taking. Most persons conceive under the term photographer only a portraitist, and only few are aware that photography is good for something else. The photographic portrait owes its popularity to its extraordinary cheapness, to its rapid mode of execution, and to its relatively greater resemblance when compared with drawings from nature. Imperfect photography can reckon, on account of these circumstances, on greater support than the drawing of a clumsy artist; and the more so as a false conceit exists that photography must have undeniably more likeness to the original—which is by no means the case, as we have seen.

Photography has driven mechanical portrait-painting out of the field—only the genuine artist is able to hold his ground against it. Portrait-photography makes greater demands than any other branch on the good taste of the photographer, his capacity to give a natural, or at least apparently natural, picturesque, unstudied pose to a person. It requires him to show the best side of the sitter, to conceal as far as possible any defect that may exist, to bring out all that is advantageous, and by a clever adaptation of light to give prominence to the essential points and leave indistinct those parts which would injure the effect of the picture. To this end the photographer is at liberty to take in the surroundings of the person, be they a chamber or a landscape, or to exclude them by screens.

In the first period of photography the pictures were commonly crammed with accessories, and incredible errors were committed with relation to position and light; but now the more advanced photographers have taken hints from artists, and latterly pictures are seen which, notwithstanding the mechanical character inherent in their production, create quite an artistic effect.

The model—that is, the person to be taken—has a very essential share in the work of photography. Not unfrequently persons go to a photographer in a very morose frame of mind, or they lose their patience through some cause or other of delay; it also often happens that people go to him with some slight *malaise*, with headache, or a restless night. This is a great mistake; the bodily or mental condition stamps itself infallibly on the picture, and often gives it a very dissimilar expression to the original, even after the photographer has used all his art upon it. In like manner, it very frequently happens that persons at the moment of being photographed put on a perfectly strange expression, force a smile, stare, let the mouth fall open, or are disturbed by the iron props which keep the head in place and are quite essential if a well-defined picture is proposed, but which are only submitted to under protest by the model, who fancies he can sit motionless without such aid.

None of these influences can be set aside by the photographer. The persons who present themselves to have their portraits taken are for the most part unknown to him. He has often only five minutes to study the faces of the persons, to find their best side, to make them pose, and to place them in accord with the surroundings. He probably attends to these matters as skilfully as any one, yet he has no power over the features of the original. Nor has he any conception if the expression of a person is his usual one, or if it be modified by ill temper or ill health. In the latter case it is impossible for the picture to please, however masterly may be the execution; yet in this case the fault lies not with the photographer, but with the original.

Another cause of failure is the inclination of many persons to choose their own position, whether of their own accord or prompted by friends. These attempts generally fail, because the errors in perspective previously enumerated are overlooked. People in general do not know this, but the photographer does. Other inconveniences result from the very nature of photography. Blue eyes are generally too light and dim, blonde hair too dark, and auburn hair reddish. Many of these difficulties are removed by clever negative retouches, but by no means all.

Still greater are the hindrances which the toilette and the changefulness of fashion furnish. Bad taste of the person is much worse and more visible in photography than in nature. Ladies deficient in taste are in the habit frequently of making their necks, which are naturally short, still shorter and thicker by necklaces. They disfigure a form, perhaps naturally excellent, by long trains; the back of their head by an ill-assorted chignon; and their hair by ribbons of the brightest and most glaring colours. In these cases the photographer can do much service by his good advice. The difficulties are even increased if groups of persons or children, and not individuals, are in question.

The children must be amused by deceptions. If the photographer wishes to succeed in taking children's portraits, he must know how to win the confidence of the little people; this is the reason why many photographers achieve great things in this sphere, and others none at all. As a child can never be long quiet, he must be taken as quickly as possible; therefore such portraits can only be taken in fine weather.

The same remark applies to groups with many persons. No atelier has twenty or thirty props at its disposal. Accordingly, the photographer is frequently reduced to the necessity of trusting to the good-will of persons in sitting still. Those groups are very ugly which show a row of persons sitting beside each other like so many pagodas. Photographers of superior refinement will occupy the attention of the sitter by some practical work, such as looking at an album, eating, drinking, or card-playing. In doing so, these persons must of necessity assume various positions, some showing their front face, others their profile, and in many cases under circumstances showing their least favoured side. In the case of groups, the photographer will attend to differences of complexion, and will find difficulties in them, and in the matter of dress. Many faces of a dark complexion have too short an exposure when others have had it long enough. But, as all must have the same length of exposure, it is not surprising that many parts of the picture appear under and others over exposed.

From these causes, no one can expect to appear to so great advantage in a group as in a separate portrait. If it happens so, this must be ascribed to chance.

It is usual for people to expect too much or too little from photographic groups. Your companion in the group is commonly well satisfied if he sees his own personality given to his fancy, quite forgetting the ungraceful arrangement of the rest, or some ungainliness among his neighbours, who do not interest him perhaps so much.

Gentlemen ought to wear dark coats. Light trousers

and white waistcoats often appear in pictures as white patches, destroying the harmony of the effect, for the principal light ought not to be concentrated on such accessories, but on the head. Ladies, in choosing their toilettes, generally overlook the abnormal working of colours. On the occasion of the triumphal entry of the Emperor and the German army into Berlin, in 1871, the young ladies chosen to grace the ceremony were afterwards photographed in their white dresses trimmed with blue, and were not a little surprised that the blue trimming was as white as the dress. Blue often becomes white in photography, though there is an exception in the case of the blue uniform of the Prussian infantry. On the other hand, yellow and buff become often black in photography; the same remark applies to red. The photographer can atone in some degree for the defect by a careful treatment of the negative in the case of clothes of uniform colour. The many-coloured toilettes, however, now in fashion, produce a disastrous effect. Materials whose beauty consists in variety of colour, it is evident, cannot make the same impression in black photography as in nature.

Persons of dark complexion, also stout persons, should prefer dark clothes. It is well known that white clothing increases in appearance the embonpoint of the figure. Thin and pale persons are advised, on the contrary, to wear light clothes, as a pale complexion would appear even paler when contrasted with black. Light clothing is always to be recommended for children. Materials should be chosen which by their lustre make a rich and picturesque impression; for example, satin, ribbed silk, taffeta, and silky materials. Woollen stuffs appear for the most part dull and lustreless, but they give very good effects in drapery. Persons of short and thick necks would do well to avoid high shirt collars, which make the neck appear still shorter. Ladies with similar attributes must lay aside velvet, ribbons, and such things around the neck; while persons of long necks will be improved by such ornaments.

The weather, the season, and the time of day present serious difficulties in photography. The days in winter being considerably shorter and darker than those in summer, commissions at Christmas are very inconvenient. Rainy days in winter are for the most part useless for photography; in summer they are clear enough. The hours immediately before and after noon are the most favourable, as we have already stated in our chapter on optics.

Besides the clearness of the weather, the amount of light admitted by the instrument is an important matter. The lighter a picture appears produced by a lens, the shorter may be the time necessary for a sitting. A lens increases strength in proportion to the size of its diameter and the smallness of its focal distance. But it is by no means possible to increase the diameter or diminish the focal distance as much as you please, for defects in the lenses, which have not yet been overcome, stand in the way of this. The strongest instruments hitherto constructed (portrait lenses) only produce small pictures of the size of cartes de visite, or at most of cabinet size. Larger pictures can be produced only by weaker instruments; therefore they require longer sittings—a circumstance that makes the taking more difficult in cloudy weather and with restless models (as children) than the preparation of smaller pictures.

Accordingly, the latter show, on the whole, a greater technical perfection. As they are also very cheap, it is natural that the small cartes de visite, introduced by Disderi at Paris, have attained a general popularity, and given rise to a new kind of album, displaying the portraits of friends instead of poetry, and almost entirely supplanting the old scrap-book.

We can only tolerate the modern album with the outlines of those we love or respect drawn by the light.

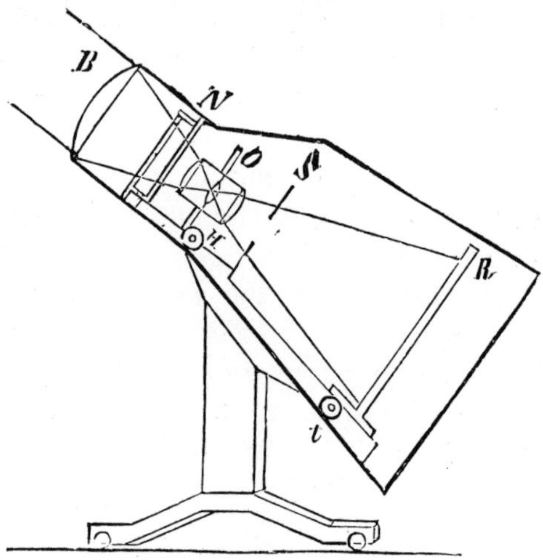

Fig. 67.

Still less can we bring ourselves to like to see cartes de visite suspended; they are too small to have any effect from the wall, and their frames are too insignificant.

Photography, like engraving, is an art that succeeds best on a small scale. Pictures of more than quarter size cannot easily be taken from nature, but life-size pictures are demanded by the public. The photographer prepares these from a small negative, by help of the magnifying apparatus previously described in the chapter on optics.

He requires for these pictures the sun, which, unfortunately, in our climate leaves him frequently in the lurch. The small negative is placed in the apparatus (Fig. 67) at *N*, and on the table at *R* a sheet of sensitive paper is stretched. Then the lens at *O* projects a magnified picture of the little negative on the screen *R*, and directly the apparatus is turned to the sun the large focal glass *B* concentrates sufficient light on the picture to occasion a rapid browning of the picture, and under favourable circumstances a life-size copy can be taken in fifteen minutes.

Much has been said of instantaneous pictures. The Deputy Faucher remarked once, in the Prussian House of Deputies, "There are now instantaneous pictures. Portraits can be stolen by this process, and it will perhaps be necessary to guard against it by the most extraordinary precautions—probably masks will have to be worn." This statement is based on a mystification. Faucher had been made the victim of one of those photographers who, by incredible boasting and by puffing themselves, seek to impose on the public. Instantaneous pictures are possible if the object is clearly illumined by the sun; therefore it is easier to prepare an instantaneous picture from a brightly illumined landscape. It is quite another matter having to do with a portrait in an atelier. Direct sunlight would produce an unpleasant glare and sharp shadows upon a dark background in a portrait under these circumstances: the eyes would have a contracted look, and a very ugly picture would be the result. As we before remarked, very powerful lenses have been constructed, which admit of shortening the time of exposure. These are, however, only suitable for very small pictures, and are only employed for small, restless objects, like children, in whose case the photographer is satisfied to get the chief part—that is, the head—as quickly as possible into the picture.

It occurs frequently that persons desire a fresh impression from a former photographic picture. We here remark that such an impression may be good, but that a photograph taken from a photograph is never so fine as an original picture. The cause of this is, on the one hand, the brown tone of the photograph, which possesses very little activity photographically; whilst, on the other hand, the paper beneath has an undue effect. This is sometimes glazed, and then produces a false light upon the reproduced picture; or it is rough, and then the veins of the paper at once cast a shade, which gives to the picture a disagreeable, coarse-grained appearance. On that account it is easy, even for unpractised eyes, to distinguish the copy from the original photograph. Such copies are frequently to be met with at fairs and in stationers' shops, and can be bought for a ridiculously small price. In most countries, however, copying from original photographs is forbidden as piracy, and this prohibition ought soon to be introduced into Germany.

It has, indeed, been remarked that piracy is advantageous to the public by making favourite pictures at a low price.* But this advantage is no compensation for

* Precisely the same argument can be advanced to defend the piracy of books, which is forbidden.

the injury thus inflicted on the author of the original photographs—he often incurring considerable expense in taking photographs in the Hartz Mountains, and the Thüringer Wald, or in making an imposing picture of a distinguished person. A great undertaking of this kind is seldom successful at first, and if his production is not protected by the law, he will prefer to give up producing such original pictures.

Section II.—Landscape Photography.

Its Scope—Difficulty of Taking Landscapes—The Photographic Tent—Significance of Landscape Photography applied to Geography—The Dry Plates—Stereoscopic Landscapes—Transparent Stereoscopes—Panoramic Pictures.

Landscape photography is a branch much less pursued than portrait photography. While portrait photography is mostly carried out on direct application, it is a very rare thing to receive orders for a landscape. These views are left to speculators, who employ photography as a means of representing favourite localities in largely frequented countries, and thereby making a profit with tourists. Thus photographers wander through the noteworthy sites of our capitals and mountains, and, as the originals are accessible to every one, each of the competitors tries to outdo his fellows by excellence of work or cheapness. The reproducer is associated with these original photographers; he does not undertake any costly journeys, but awaits the issue of original photographs to copy them at once, and offer them at a low price. The inclination of the public is here favourable to the cheap seller. A landscape is seldom bought for its value as a work of art, but more as a souvenir of a happy hour, or as a reminiscence which in subsequent years will recall some interesting object, whether a statue or a castle; therefore less is demanded in the case of landscape and architecture views, and this is the reason why landscape photography is not at present in a very high state of perfection. The English are relatively the best in this branch, because they ask good prices for their pictures and are protected against piracy. The Swiss views of Mr. England have a universal celebrity in Germany; pictures of equal merit are only produced by Baldi and Würthle at Salzburg. Braun of Dornach also deserves an honourable mention, having produced excellent landscapes, his Swiss views being known everywhere.

Superficial observers entertain the belief that landscape photographs must be as good as others, as the object remains always the same, and all are prepared by the same process.

Both assumptions are, however, erroneous. The object is not always the same, for a landscape appears under very different aspects in the morning and evening light, or in fine and clear weather. Whoever studies these effects of light will soon discover at what hour a landscape will look most beautiful, and will choose it for taking his view. Accordingly, his picture will surpass greatly that of a superficial and hasty photographer, who takes the landscape as he finds it. The choice of the position is equally important. By comparing pictures taken at different heights (see page 147), it is evident that the whole scene in many landscapes changes by standing a little higher or lower, or a few more paces to the right or to the left. The man having the eye of an artist, who knows how to seek the best position, will at all times give the best picture.

The same remark applies to taking architecture and sculpture. It is evident that a photographer who undertakes this must be partially favoured by wind and weather. A breeze stirring the trees often injures his view, which may be impeded the whole day by wind and rain. To this difficulty may be also allied that of an unmannerly class of people, who insist on being taken with the picture, and thrust themselves full in front of the photographic apparatus, making the attempt impossible—a weakness that is more commonly met with in Germany than elsewhere, and is the more inexplicable as they generally never see anything more of the picture.

It is a great inconvenience for the landscape photographer that all the chemicals, cups, bottles, glasses, which are requisite for the process, must be carried on the journey; nay, more, the photographer needs a transportable dark chamber in which he can prepare his sensitive plates.

The accompanying figure represents an apparatus of this kind, together with the photographer. It is only the upper part of his body that is in the tent, but the intermediate space is impervious to light through the curtain. For the sake of facility of transport, everything in a dark tent of this kind is contracted into the smallest space. A yellow glass *q* lets light into the interior. The silver bath is in a box *y*, and the necessary water is in the cistern *x*, from which a pipe passes into the interior. The whole tent can be folded up, and forms a box of the size of *z* in the figure.

Though these arrangements are very compact and compendious, they are still of considerable weight, making the ascent of difficult places, such as the Finsteraarhorn, the Wetterhorn, and the Jungfrau, impossible.

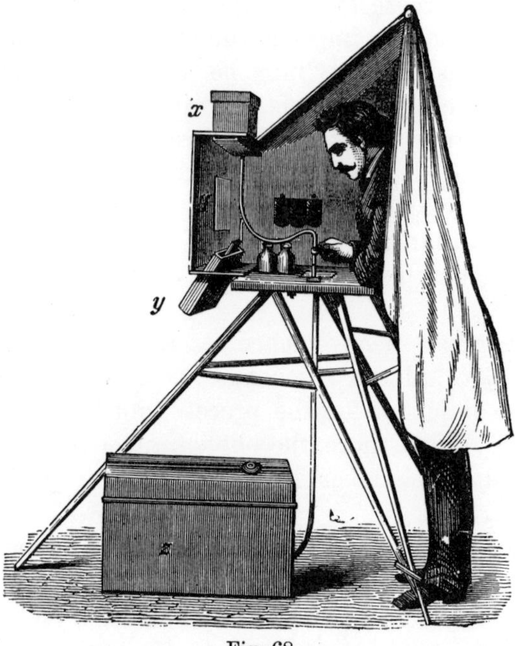

Fig. 68.

Dry plates are important to produce views of this kind, as they can be prepared at home and taken on a journey. They dispense with the dark tent, collodion, silver bath, and the water for washing. The dry plate and photographic apparatus suffice in this case. We have already described the production of dry plates, and how they are prepared by washing an ordinary plate of sensitized collodion, pouring on it some substance absorbing iodine, such as tannin, and then leaving it to dry. Unfortunately, plates prepared in this manner are less sensitive than the fresh wet plates, and the pictures they give appear less fine than those taken with the freshly prepared plates; the results, therefore, cannot be depended upon. Many plates spoil after a lapse of time; then the view obtained cannot be judged till it has been touched up at home, and the result is often very imperfect, nor can it be corrected far from the place where it was taken. For these reasons the moist process has been preferred in taking landscapes, notwithstanding its inconvenience, and only a few photographers work with dry plates.

Stereoscopic views are very popular in the landscape branch. Though so limited in size, they present landscapes in so plastic a form that they distance even larger pictures in their effect. We have already described how these views are taken. If the light is bright and the lens large, instantaneous views can be taken with the stereoscopic apparatus, and have been offered largely for sale.

These transparent stereoscopic views on glass, prepared by Ferrier and Soulier, are wonderfully beautiful. They are produced on a collodion film, by placing the glass negative taken from nature on a dry plate, and then exposing it. It then copies the negative exactly in the same manner upon the sensitized glass plate, or upon the sensitized paper, only the invisible light impression must be first developed by the application of pyrogallic acid. As the production of such glass positives requires a more minute and lengthened treatment than the paper pictures, their price is higher.

Latterly, however, by the help of a printing process (the Woodbury process), it has become possible to produce these glass pictures at a considerably cheaper rate. We refer to this process further on.

Though at first sight landscape photography may appear unimportant, yet it is of the greatest use for geographical information. There is no better medium for conveying a true picture of foreign lands, of rocks, plants, and animal forms, than photography. It has even become an essential auxiliary in exploring expeditions, being alone capable of giving a perfectly faithful description of what has been seen. I admit that the inconvenience of transporting a photographic apparatus, and the injury to which the chemicals are easily exposed, limit the use of photography in exploring expeditions, and require a very expert photographer; but that these obstacles can be overcome is proved, among others, by the excellent views taken by Count Wilzek and Burger at Nova-Zembla, Baron Stillfried in Japan, Burger and Lyons in India, and Dr. G. Fritzsch in South Africa. We shall show in the following chapter the importance of landscape photographs for land surveying.

Panoramic views are quite a special branch of landscape photography. The noted photographer Braun, of Dornach (Alsace), offered for many years pictures for sale which contained half the circumference of the panoramas of the Rigi, of the Faulhorn, of Pilatus, and other well-known points. It is evidently impossible for a fixed camera to command at once such a panorama; nor can the human eye do this, for the most we can survey is 90°, and this is only a quarter of a circumference. If we wish to see the whole circumference, we are obliged to turn round. Martens, a German engraver residing at Paris, conceived the idea of taking panoramic views with the help of a revolving camera, or of a revolving lens in a camera. Let the reader imagine a camera with a cylindrical hinder surface *p p* (Fig. 69)

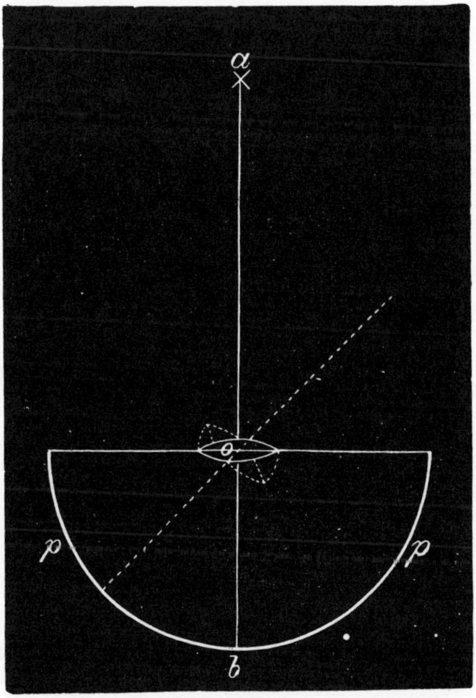

Fig. 69.

represented in outline, also a lens *o*. Then the image of any point *a* is situated on the line *a o b*, which is drawn from *a* through the centre of the objective. If the lens revolves round its centre, the image remains immovably at its place *b*; if it were to revolve to any other point than its centre, the image would be displaced. It is therefore evident that, if the lens revolves round its centre, it can imprint successively an image of half the horizon on the cylindrical surface. The only problem then, therefore, is to produce a sensitive cylindrical surface. This is not difficult with sensitive paper, but much more difficult with glass, which is extremely brittle in this form. Accordingly, Brandon introduced a smooth plate, which rolls as it were round the cylinder *p p*; that is, which during the revolution of the lens is moved in such a fashion that it always remains perpendicular to the axis *o b* of the lens. The mechanism of such a camera is rather complicated, but it has maintained its ground in practice, and numerous panoramic views have been taken with it. We must confine ourselves here to cursory remarks; those who seek for further details are referred to Vogel's "Manual of Photography."

SECTION III.—PHOTOGRAMMETRY, OR LEVELLING BY PHOTOGRAPHY.

Relation between Photography and Measurement—Principle of Trigonometrical Measurement—Projection of Maps—Photographic Measurement of Altitudes.

An essential difference between a photographic view and an artist's painting is the fact of its not being the production of the operator's will, but that its outline and design are subject to determinate laws. All photographic views are produced by means of lenses. A lens view of this kind is always an exact central perspective; that is, each point of view lies on the straight line which can be drawn through the optical centre of the lens. Let *a b c* (Fig 70) be three objects in nature, *K* a camera (of which the outline is given to facilitate comprehension), and *l* its lens. Then the images of the different objects are situated on the produced short lines *a o*, *b o*, *c o*—that is to say, on *a′ b′ c′*; therefore they have in the picture exactly the same relative position as in nature. Accordingly, a good photograph can serve to determine accurately the position of objects in nature; that is, to construct maps of the piece of ground that has been taken in the view.

For example, let the reader conceive an image, which stands upright in the camera of the diagram annexed, brought down flat upon the paper. Then in the middle of the field of view, at the tree *b′*, let a perpendicular line be drawn equal to the focal distance *o b′*. It is only necessary, after this, following the figure, to construct the lines *c′ o* and *a′ o F′ o*, in order at once to find the directions in which the tower, the flag, and the trees will be seen from the position *P*. If now a second view be taken from a point *P′* which lies in the direction of the flag *F*, a second view is obtained *c″ b″ a″*, which naturally looks quite different from the first in consequence of the change of position. If this view from the second position be also brought down to it, and a line *b″ o*, equal in length to the focal distance, be drawn to the second position, then the lines *c″ o* and *a″ o* give again the direction of the lines from *a b c*. If these lines be sufficiently produced on the paper, they intersect at points the situation of which corresponds exactly to that of the object; and thus, in two views at two points, the means is afforded of constructing a map in which the situation of all points contained in both views is exactly given.

A different procedure is followed in ordinary trigonometry. In that science, the first step would be to measure the distance *P P′*, then to set up an instrument for taking angles at *P*, and to determine the angle made by the line *P P′* with the lines *a o*, *b o*, *c o*; the same

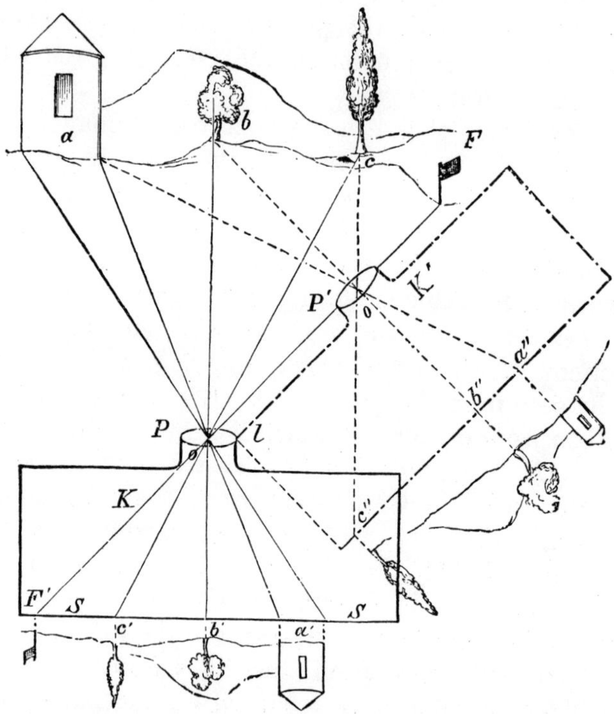

Fig. 70.

operation being repeated at the other end of what is called the station line *P P′*. It is evident that as many measurements must be made at both points as there are objects of interest, whereas a photographic view taken once for all fixes all objects correctly in their relative positions. Accordingly, there is a considerable economy of time in applying photography; and this is of great moment in war, when frequently, in consequence of interruptions on the part of the enemy, the leisure is wanted which is necessary for triangulation; also in journeys, when the stay at particular places is too short to make observations requiring time.

Therefore this process has great advantages in exploring expoditions; and landscapes taken photographically have a twofold value: not only do they give a view of the country, but also data for the projection of maps. I admit that to this end two views are necessary, which must be taken from the end of a station line. Then the taking of these views must be carried out with mathematical accuracy: the camera must be placed in a perfectly horizontal position; its lens must give a perfectly correct image; the plates must be absolutely level, etc. But all these conditions are not easily obtained. To this other difficulties are added, proceeding from the very nature of photography. This art requires clear, bright weather; with a troubled sky, or when the atmosphere is veiled—the aerial perspective of landscape painters—it often gives remote objects so indistinctly in the view that no correct measurement can be made of them, though the surveyor can clearly distinguish and measure from nature in such weather. Further, the direct action of the sun offers difficulties to photography. If it stands in front of the camera—that is, if it shines full on the objective—it often occasions fogging on the plate, greatly modifying the value of the view for purposes of measurement. All these circumstances militate against the application of photogrammetry, as this mode of measurement has been called by Meydenbauer, who has long used it. Meydenbauer prepared a good map of the

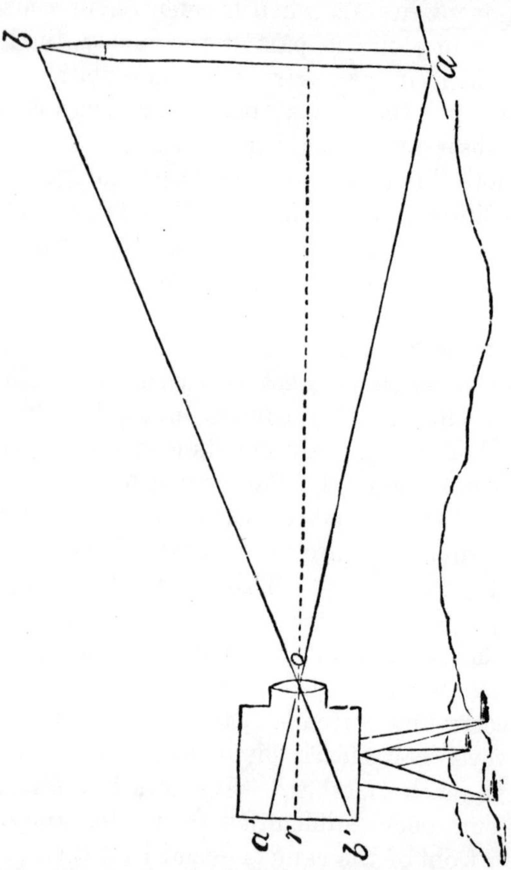

Fig. 71.

Unstrutthal according to this method. But the experiences during the campaign of 1870 are not so satisfactory (the royal Prussian staff tried the process before Strasburg)—perhaps the imperfection of the apparatus occasioned the unsatisfactory results. It is to be hoped that future attempts will succeed in making this important method practically available in the interests of geography.

Photography can determine the elevation of mountains and of buildings, as well as determine positions in a plain. Let it be supposed that *a b* (Fig. 71) is a tower, represented in the accompanying diagram as imprinting the image *a′ b′* on the photographic apparatus. It is evident that the image will be smaller than the object. According to a well-known mathematical principle, the magnitude of the image *a′ b′* to that of the tower *a b* is as the distance of the image from the objective *o r* to the distance of the tower from the objective. This gives the proportion:

$$o\ r : E = a'\ b' : x,$$

in which by E is understood the distance of the tower from the camera, which can be measured. The height of the tower can be easily found from the above proportion.

Meydenbauer has deduced the dimensions of the ground-plan and elevation of a house from its photograph.

Section IV.—Astronomical Photography.

Its Application—The Photographic Telescope—Taking Eclipses—Protuberances—Corona—Sun-spots—Enlarged Images of the Sun—Rutherford's Labours—Astral-Photography—Pictures of the Moon—Spectral Photography—Photography and the Transit of Venus.

The province of astronomical photography may embrace two kinds of operations: first, it has to give a faithful representation of the phenomena of the heavens—which change so rapidly that the operator cannot follow them; for example, the phenomena of eclipses, or others which are inconvenient to represent, such as sun-spots. Secondly, astronomical photography has to produce views of heavenly bodies that can be used for measurements. Photography has made successful attempts in both these walks, and it is employed daily as an auxiliary to produce views of sun-spots at several observatories; for instance, in Germany, at the observatory of Herr Von Bülow, Privy Councillor, at Bothkamp, near Kiel.

The art and mode of preparing astronomical pictures differs little from that of ordinary photographs. An ordinary photographic apparatus can be used for this

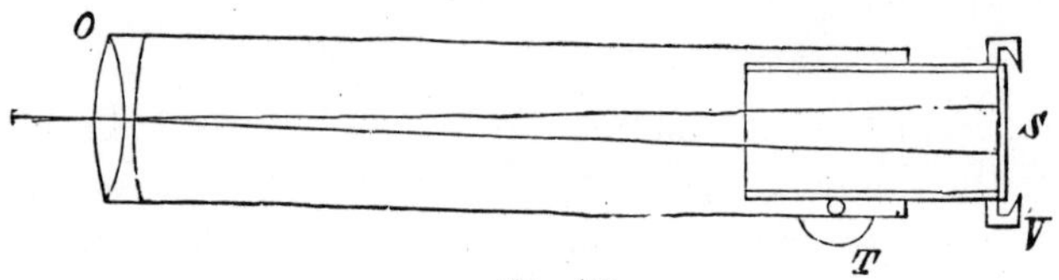

Fig. 72.

purpose, were it not that it gives too minute views of very remote objects, as the stars. The size of the picture bears a direct relation to the focal distance of the lens; therefore, in taking astronomic photographs, astronomic lenses are used where focal distance is very long, by converting an astronomic telescope into a photographic instrument.

The accompanying figure shows a telescope of this kind adapted to photographic purposes. The objective *O* remains in its place, the eye-piece, which is fixed at the other end of the tube, is taken away, and an apparatus *V* (Fig. 72) is substituted for it, which is identical with the hinder part of a photographic camera

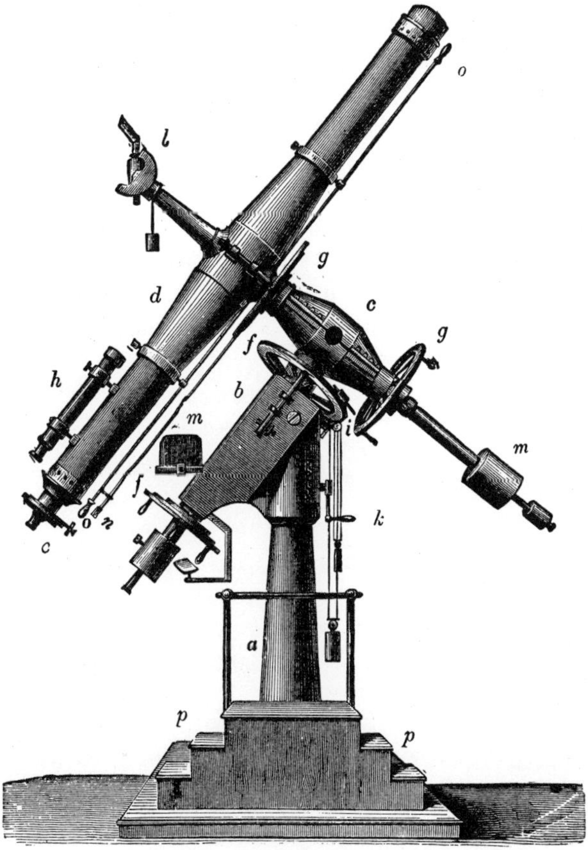

Fig. 73.

Thus it contains a ground glass slide *S*, which, after this apparatus has been firmly fitted in, can be exchanged for a sensitive plate. This firm fitting is effected by moving the spring *T*.

But now a difficulty occurs in the movement of stars, which leads to the necessity of the telescope following this movement if the views are to be sharply brought out. To this end the stand of the telescope is furnished with a clockwork *p p* that causes it to revolve in the direction of the course of the stars, so that the telescope is what is called parallactically set up. Fig. 73 shows an arrangement of this kind.

The oblique leg of the telescope resting on the foot *a* is parallel to the earth's axis. On this stand the polar axis of the telescope revolves with the hour circle *f i*, by the working of the clockwork, once every twenty-four hours.

The telescope *d d* is not fixed immediately on the polar axis, but on an axis *c*, perpendicular to it; it can be turned round the latter (the declination axis) in all directions perpendicular to the axis *c i*. It is only the movement of the two axes that allows any star you please to be brought into the field of view of the telescope.

The first attempt to employ photography for astronomic views was made by Berkowsky, at the Royal Observatory, in the year 1851, by the help of Bessel's noted heliometer, during a total eclipse of the sun. He obtained a dageurreotype, the beauty of which was much lauded, and which showed very well the remarkable phenomena that appear during an eclipse of the sun—flame-like formations that stand out in the darkened disk of the sun, being what are called protuberances. In the year 1860, Warren de la Rue in England, and Secchi at

Rome, undertook an expedition to Rivabellosa, in Spain, to observe the eclipse of the sun, and both produced interesting views on collodion plates. In 1868, the Government of the North German Confederation equipped an expedition to observe the eclipse of August 18th, and sent Dr. Fritzsch, Messrs. Zencker, Tiele, and the author of this work to take photographs. Another photographic expedition was sent out by the English Government to India. Besides these, the German, English, Austrian, and French Governments sent out expeditions for the ocular observation of the phenomenon.

Obstacles were, no doubt, encountered by these expeditions, nevertheless they produced results that finally settled the question about the nature of the protuberances, and moreover gained experience that materially lessened the labour of subsequent photographic observers.

We proceed to introduce a description of the expedition to Aden, giving a faithful account of the obstacles associated with an undertaking so simple in appearance. The author wrote from Aden the following account of his arrival and residence at that place :—

"The aspect of Aden is by no means cheerful. An almost entirely bald, savage, and broken mass of rock, the remains of an extinct volcano; in front and in the midst warehouses, shops, coal-sheds, flag-staffs, etc. ;— such was the appearance of the place that was to be our residence for a fortnight. The colour green was entirely wanting in nature.

"Our luggage and ourselves were conveyed to land amid the shouts, quarrelling, and tumult of the Arab mob. On landing we learned that our colleagues, who had preceded us, had been received in the most courteous manner by the British authorities, and that two Indian huts—bungalows, usual in this climate—on the east side of the peninsula, had been assigned us as our station.

"After a long search we found the locality and our comrades, together with the members of the Austrian expedition, Dr. Weiss and Messrs. Oppolzer and Riha, and in as excellent quarters as could be wished on this desert coast. The English authorities acted the part of host in the most generous manner. A whole staff of servants, cook, etc., waited upon us; carriages, camels, and donkeys were at our disposal, and all our wishes were gratified or anticipated. Thus our bodily comfort had little to desire; the temperature (26° Reaumur), might be called low for the Red Sea, for a fresh breeze was always blowing on the heights of the Marshagill, on which our bungalows stood, and contributed greatly in refreshing the air.

"There still remained ten days to prepare for taking views of the eclipse. They were employed in preparing stands for our telescopes, in putting them up, and in setting them in order. We used as observatory a bungalow, which we partly unroofed, in order to look through the roof with the telescope, and we converted the rest of its interior into a laboratory, washing-room, and store-room. In this telescope cage—for it was nothing more—we were tolerably protected from the wind, but less so from the dust. Water was brought to us by donkeys, in leather bags. Two tents that we had brought from Europe answered the purpose of dark chambers. Spare apparatus for taking landscapes and portraits, that we had brought with us, gave us the material for taking views of the country and its population, and were also a useful means of testing our chemicals.

"Some trifling defects in the latter were quickly remedied, but it was not so easy to remove the effects of the dust and human exhalations. During the slightest exertion in that damp atmosphere, the perspiration flowed from the body in streams: it ran from the tips of the fingers, dropped from the face, and often a well-cleansed or prepared plate was spoilt by a drop of sweat. Nevertheless, practice taught us how to encounter this obstacle; some attempts at taking the sun, etc., turned out successful; we were able to look on tranquilly to the eclipse. Only one thing gave us serious uneasiness,—that was the weather. All accounts of Aden had unanimously represented its sky as perfectly clear, competent witnesses having asseverated that it rained there at most three times a year, and that clouds were exceptions.

"We were therefore not a little surprised when, on our arrival, we found the volcanic heights of Aden quite concealed with clouds, and when we were greeted with a shower of rain on the following morning. But we became still more anxious when, day after day, the sun was concealed by clouds, and this weather became worse rather than better in the course of time. The prospect of succeeding in our main object looked dreary enough, and soon all our hopes deserted us.

"On the day of the eclipse we left our quarters about 4 a.m. Nine-tenths of the sky were cloudy. We set to work in a resigned frame of mind. It was the undertaking of the North German expedition to photograph the eclipse throughout its continuance. For this purpose we used a telescope with a six inch lens without focal difference, with a focal distance of six feet. This lens, made by Steinheil, gave an image of the sun three-fourths of an inch in diameter, which could be taken on a photographic plate by the help of an ordinary box with slides for two views. As the sun and moon move, such an instrument, if stationary, would only give ill-defined views. Accordingly, the telescope was connected with wheelworks that gave it a movement corresponding with that of the heavenly bodies. To avoid all agitation of the telescope, the closing lid of the objective was not placed close to the telescope, but at a separate stand, and was connected with the telescope by an elastic hood.

"The duration of the total eclipse was at Aden three minutes, in India five. We had, however, chosen our station at Aden because photographic observers were already present in India, and because the eclipse began first at Aden (about an hour sooner than in India). Thus, by comparing our observations with those in India, a judgment might be formed whether those wonderful phenomena of light called protuberances, during a total eclipse, changed or did not change in the course of the eclipse. It was our present endeavour to take as many views as possible of the phenomenon in three minutes. To this end we had regularly practised, as artillerymen do with their cannon.

"Dr. Fritzsch prepared the plates in the first tent, Dr. Zencker pushed the boxes into the telescope, Dr. Tiele exposed them and developed them in the second tent.

"We had determined that it was possible in this manner to take six views in three minutes.

"The decisive moment approached. The cloudy sky, anxiously surveyed by us, showed to our great satisfaction a few breaks through which the disc of the sun, partly concealed by the moon, and appearing as a crescent, was visible. The landscape appeared in the strangest light, being almost a half-and-half mixture between sunlight and moonlight. The chemical influence of light showed itself very weak. An experimental plate only gave a view of the clouds after fifteen seconds' exposure. The sun's crescent became gradually smaller, the break in the clouds gradually increased, and we took heart.

"The last minutes preceding the total eclipse, which occurred at 6.20, fled on wings. Dr. Fritzsch and I crept hastily into our tent and remained there, preparing plates and developing. The consequence was that neither of us saw anything of the total eclipse. Our labour begun, the first plate was exposed, as an experiment, from five to ten seconds, in order to see what was the proper time for exposure.

"Mohammed, our dusky attendant, brought the first box into the tent to me. I poured the iron developer upon the plate, waiting breathlessly to see the result. At this moment my lamp went out. 'Light! Light!' I exclaimed; but no one heard me—every one had enough to do. I stretched my right hand out of the tent, holding the plate with the left, and fortunately grasped a small oil lamp, which I had placed ready for all emergencies, and now I saw the image of the sun appear upon the plate. The dark rim of the sun was surrounded by a series of peculiar prominences on one side, while on the other side appeared a singular horn,—both phenomena perfectly analogous in both views. My delight was great, but there was no time for rejoicing ; soon the second plate, and, a minute later, the third plate were in my tent. 'The sun is emerging,' exclaimed Zencker. The total eclipse was over ; but all this appeared as the work of a moment, so quickly had the time passed. The second plate showed under development only faint traces of the view, a passing veil of clouds had almost destroyed the photographic operation at the moment of exposure. The third plate showed again two successful views with protuberances on the outer rim.

"Rejoicing in this success, the plates were washed, fixed, varnished,—no doubt with very imperfect materials,—some copies were taken on glass, and these, to obviate loss, were transmitted separately to Europe.

"Our extraordinary good fortune is apparent from the fact that, at a place distant only half a league from our station, nothing was seen of the total eclipse on account of the veil of clouds.

"We did not stay long at Aden after our chief object had been attained: in three days the steamer proceeded to Suez. Telescope, wheelwork, and our heap of instruments and chemicals were quickly packed, placed on camels and conveyed to the harbour. On the 21st of August we bade adieu to the barren, rocky island, and started for Suez."

Aden was one of the points where the eclipse was soonest seen. As previously stated, the English had also equipped a photographic expedition, which stationed itself at Guntoor in India. The eclipse was observed an hour later in India than at Aden. The same protuberances appeared in the Indian photographs as in those at Aden, but they present a very different form, which seems to show that these prominences are not compact bodies, but formations of a cloud-like nature; and this supposition was converted into certainty by Jansen's observations with the spectrum, made simultaneously. Jansen showed that in a total eclipse the protuberances display clear lines in the spectroscope; but, as this only takes place with gaseous bodies, the question about the nature of the protuberances was solved. Jansen determined at the same time the exact position of the clear lines of the spectrum, and detected the nature of the gaseous substance as glowing hot hydrogen. He subsequently made the discovery that an eclipse was by no means necessary in order to detect the clear lines of the protuberances. They are seen on clear days, if the eyepiece of a spectroscope be directed to the sun's rim, and the changeable nature of these protuberances can be observed on the appearance and disappearance of these clear lines. Zöllner of Leipzig even detected this sudden flaming up through the spectroscope, also the sudden breaking away of gas clouds from their substratum, and their dispersion, all in the space of a few minutes.

We add here a faithful copy of the Aden photographs, which we have taken from Herr Schellen's excellent work on spectral analysis, published by Westermann at Brunswick. The first view gives us the eastern rim of the sun; the western was covered by clouds. It is easy to recognize in it the large horn-like protuberance, which has an elevation of 184,000 miles, and gives an idea with what immense force masses of gas are projected over the surface of the sun. It shows, further, the remarkable protuberance to the left, in which the masses of gas appear like powerful jets of flame driven sideways by a tempestuous wind; a light region

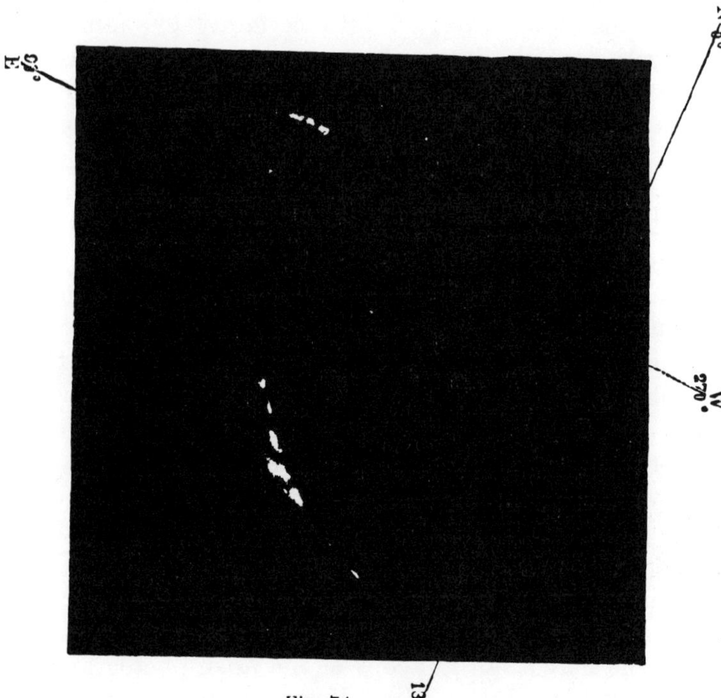

Fig. 74.

surrounding the protuberances forms the glowing hot stratum of vapour permanently surrounding the rim, and is named chromosphere.

The second view presents only a series of point-like protuberances on the western rim of the sun, but these points are so large that our earth could almost find room in them. The eastern part of the sun was under the clouds during the taking of this view.

Finally, the third view gives a perfect representation of the total eclipse as it was observed in India. Besides the protuberances seen at Aden, there is another on the

Fig. 75.

western rim of the sun, which was quite covered by clouds at Aden.

Photography has been latterly applied to the observation of total eclipses on a more magnificent scale. Thus, on the 7th of August, 1869, hundreds of photographers were actively employed in observing the total eclipse of the sun at Iowa, in North America, and more than thirty telescopes were set up to fix the phenomenon. By these observations, the question respecting the nature of the protuberances was finally set at rest, and the only question that remained related to the corona. By

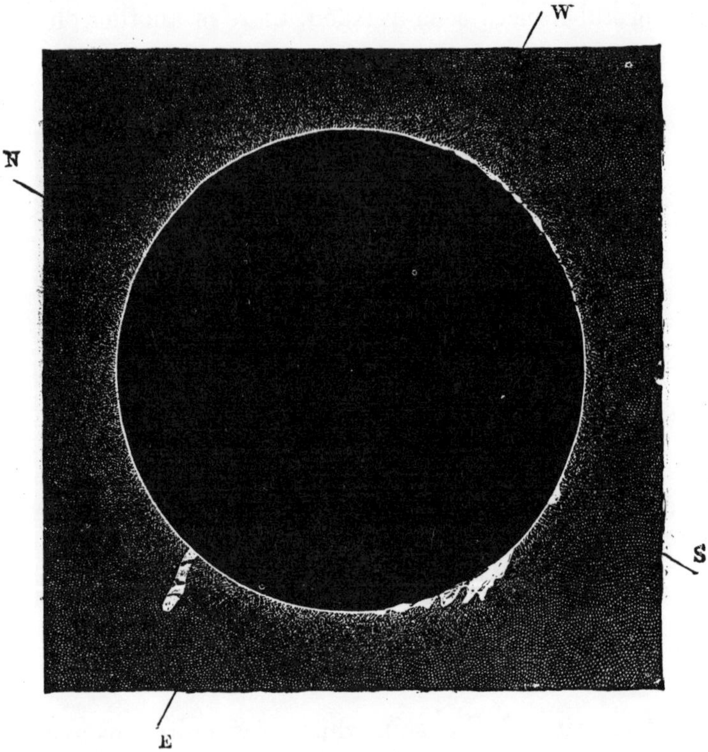

Fig. 16.

corona is implied a kind of nimbus of white light encompassing the sun when totally eclipsed. Many observations of total eclipses have been undertaken for the solution of this question. A very beautiful view of the corona was obtained by Whipple, at Shelbyville in Kentucky, August 7th, 1869. A much longer exposure is required in the case of the corona than in taking the protuberances, on account of the feeble light attending the phenomenon. At Shelbyville, the exposure for the corona lasted forty-two seconds, whereas five seconds

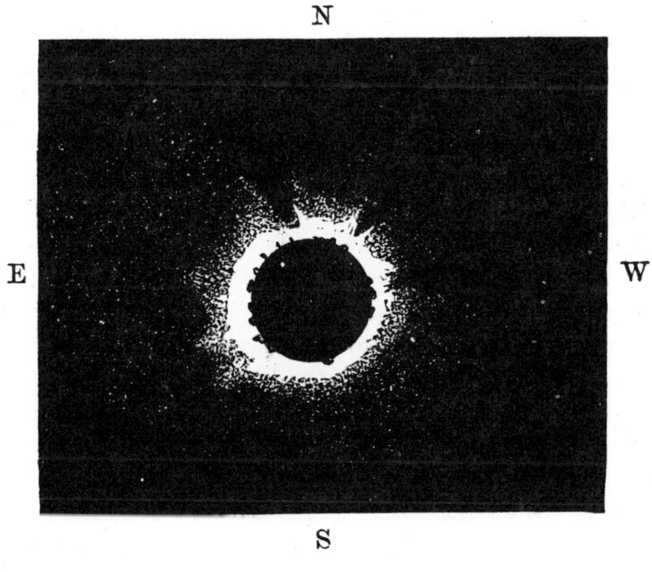

Fig. 77.

sufficed to take the protuberances. Nor was the nature of the corona as yet determined.

In 1870 the English sent out an expedition for the observation of the corona, conducted by Lockyer, to Catania, and the author accompanied it. Unfortunately, owing to the unfavourable weather, the observations were only partially successful. Nevertheless, a detachment of the expedition, conducted by Brother, to Syracuse, succeeded in obtaining a good view of the corona, and we give a faithful woodcut copied from this photograph.

The black prominences round the sun's disc give the situation of the protuberances which were visible on the day of the eclipse. But we lay stress on the fact that they are not visible in the photograph of the corona. To take a view of the corona requires an exposure eight times as long as the protuberances. During this long exposure the protuberances in the view received too much influence, and are therefore paler, so that their outline becomes confounded with the indistinct parts.

Photography is applied to other important objects besides eclipses. Views of the sun are taken daily with it. The observation of centuries has established that the sun is continually changing: spots appear, increase, and disappear. All these phenomena were at an earlier date explained as openings in the cloudy luminous atmosphere of the sun, which was supposed to surround its dark central mass. Now they are looked upon as immense whirlwinds, which rage in the atmosphere of the sun (see Schellen's "Spectral Analysis" page 200), or as cloud-like condensations. Their nature has not been perfectly ascertained. These sun-spots follow the revolution of the sun's body round its axis, and experience manifold changes during this time. It has been only by means of these spots that the duration of the sun's revolution has been determined. Recent observations have established that the size and varying number of the spots change periodically, and that these are connected with the magnetic phenomena of our earth.

These circumstances have led to a more devoted study of the spot formations, and photography has offered a valuable aid to them. It gives at a particular moment a faithful view of the sun's surface, and photographs taken daily give us the most exact representation of its spots, their size and number; and a comparison of the views during one month gives an instructive survey of the changes on the sun's surface, as they relate more faithfully than words the history of the central body of our planetary system. The amateur Lewis Rutherford, at New York, who has made valuable contributions to

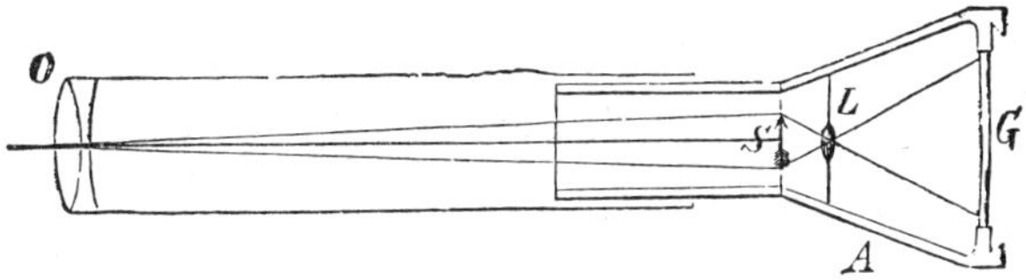

Fig. 78.

astronomical photography, has taken a great number of these views at the photographic observatory built at his own expense.

These views, taken on successive days, exhibit manifold groups of spots, often of considerable size; and the change in their form and position is accurately discerned (this change consequent upon the revolution of the sun's body). These impressions are not prepared, as were the pictures of the eclipse, in the principal focus of the telescope, but in an appendage (Fig. 78) which answers the purpose of a magnifying apparatus. This contains a small lens *L*, which projects on the ground glass *G* an enlarged image of the small representation of the sun *S* produced by the great lens *O*.

In this manner Rutherford obtained immediately a view of the sun of about two inches diameter. This enlarging apparatus is not to be recommended in the case of eclipses of the sun, for the clearness of the optical image produced by the great telescopic lens is materially weakened by the enlargement. When the enlargement is twice as great, the weakening of the light is fourfold; when it is three times as great, the weakening is ninefold, and so on. In views of the unclouded sun this is of no consequence, for its light is so intense that it bears a considerable enlargement, and yet remains clear enough to give a view on a momentary exposure. But it is otherwise with the protuberances, which give out much less light, and which, on the application of an enlarging system, would produce such faint views that a longer exposure would be required than the duration of the eclipse.

The solution of other important astronomical problems has been attempted with the help of photography; for example, the production of views of the starry heavens.

The object of these views of the stars is a representation of the constellations, or the relative position of the stars. It was always one of the principal objects of astronomy to determine the position of the fixed stars. It may be thought that the catalogue of the stars is already completed, and that the matter is settled; but this is not the case. As far as photography can at present be applied, that is, to stars of the ninth magnitude, the catalogue is not complete. Moreover, the measurements of the past may require correction in consequence of improved methods.

The photographic process has a scientific importance for this end, because it offers advantages in the facility and correctness of its results. Many readers may inquire why we take so much trouble to discover with the greatest accuracy the positions of thousands and millions of fixed stars. The answer is that the fixed stars are not stationary, as their name implies; nothing is stationary and at rest in nature, and hence their study is never at an end. No doubt the fixed stars change their position so slowly that the builders of the pyramids four thousand years ago beheld the constellations much as we do. It is only the minutest astronomical measurements that show such a change within a limited number of years. However, the study of the proper movement of the fixed stars has now begun, and requires very accurate measurements carried on for generations.

Another interesting point comes into consideration in this connection. On the one hand, the fixed stars are not without movement; on the other, their distances from the earth vary, and those of the nearest are immensely great. The photographer who wishes to have a graphic view of an object, will always seek to take it from different points. Two views of a moderately remote object, taken from two points that are only two inches apart, appear different to the eye, and produce, when viewed in a special manner, a stereoscopic effect. No distance on earth is great enough to give different pictures of the same constellation; nevertheless, within the space of one half-year we describe a circle round the sun having a diameter of 184 millions of miles, so that in half a year we are 184 millions of miles from our present position. This enormous distance is in certain cases sufficient to show a change in the mutual position of certain stars, though the distance is insufficient for the naked eye and the stereoscope, and only available for the finest astronomical instruments. By this means the distance of the nearest fixed stars has been determined, amounting to billions of miles.

By careful comparative measurements of positions of neighbouring stars, continued for years and centuries, a change can be proved to exist, and the proper movement of the stars can be calculated. The distance of the stars can be deduced by a careful collating of the yearly recurring changes in the positions of the stars. It is apparent that the fixing of these positions by photography, which admits the taking of measurements at any given time, must be of the greatest value for both these astronomical problems.

The photographing of the stars was first introduced into science, about twenty years ago, by Professor Bond, of Cambridge, Massachusetts, but it was Mr. Lewis Rutherford, of New York, who perfected this method. He constructed a photographic objective of 11 inches diameter and about 13 feet focus. This objective shows a considerable focal difference; that is, the violet and blue rays have a different focus from the yellow and red. If a clear image of the star is taken, the sensitive plate is adjusted to the focus of the yellow rays, and the chemically operative blue rays are then situated outside the sensitive plate, and produce a faint impression. The plate must therefore be adjusted to the focus of the blue rays; but this is not so easy to find. After it has been found approximatively, it is corrected by taking different photographs of the star, changing the position of the plate. The point is determined from which the best and sharpest view is taken, and, by continual repetitions of the attempts, the chemically operative focus of the lens of 13 feet focal distance can be accurately determined to within $\frac{1}{150}$ of an inch. It is well known that all heavenly bodies have the same focus, on account of their great distance. No photographic objective gives a picture with a large surface perfectly correct. Accordingly, with the accuracy required by astronomic photography, the surface to be devoted to the image can only be very small, or about $1\frac{1}{2}$ degrees. Errors in drawing, which must appear here, are controlled and corrected by photographing a very correct scale, and comparing the picture with the original. A field of $1\frac{1}{2}$ degrees, or three times the moon's diameter, embraces the well-known image of the Pleiades.

The telescope of Rutherford is arranged as in Fig. 73 (p. 173); it is moved by clockwork, to be able exactly to follow the movement of the stars.

The views of large stars taken with it, after a short exposure, all appear like small round points that can only be seen through the magnifying glass. In the case of a long exposure their size depends, fundamentally, on the more or less strong vibrations of the atmosphere, which occasion the flickering of the stars. Stars of the ninth magnitude can be photographed with an exposure of eight minutes; these stars are ten times weaker than the faintest that can be detected on a clear night by the naked eye, and their images are very small points. It would be difficult to distinguish these small points from dirt spots on the plate. To do this,

Rutherford makes use of an ingenious process. He brings the telescope, after the first exposure of eight minutes, into a slightly different direction, and makes another exposure of eight minutes, while the clockwork continues to operate, and moves the telescope correctly in this second direction.

In this manner two images are obtained of every star on the plate, closely adjacent; the distance and relative position being in all the same. These double views can be easily found on the plate and distinguished from spots. If the telescope stops, it is evident that the images of the stars make a movement on the plate, the bright stars describing a line. This line is of great importance to determine the direction from east to west on the plate. For faint stars which leave no line a third exposure is necessary, to determine this direction; the same thing takes place after the clockwork of the telescope has been stopped for some minutes.

Rutherford has already taken numerous views of the stars, and they will serve as important means of comparison, after the lapse of centuries, in order to discover what change has taken place in the position of the fixed stars.*

But another heavenly body invites us specially to study it by the help of photography; that is our neighbouring satellite, the moon. The unassisted eye recognizes its uneven surface ("mountains in the moon") and the varying shades of its ground (moon spots). Its surface contains a thousand problems, appearing as a rigid, almost vitreous, waterless, airless waste.

* Details respecting Rutherford's observatory are contained in the "Photographischen Mittheilungen," Jahrg. 1870. Berlin: Oppenheim.

Warren de la Rue tried to take this singular globe, which is so near to our earth and yet so different; he actually prepared a small view of the moon with the help of a telescope, which he enlarged to 24 inches with the aid of an enlarging apparatus (p. 96).

The moon gives out less light than the sun. It is therefore taken to the best advantage in the principal focus of the telescope. (See Fig. 72, p. 172.) In the most favourable case, three-quarters of a second suffices for exposure, but it is rare to obtain sharp negatives, owing to the disturbance of the atmosphere. Therefore, to obtain a sharp image of the moon is a test of patience. After Warren de la Rue, Rutherford obtained notoriety by taking moon-pictures; his improved telescope, set up purposely for photographic purposes, gave a still sharper image of the moon than De la Rue's, and our frontispiece is a diminished copy of the enlarged picture of the moon according to the original, for which we are indebted to Rutherford, forming a genuine map of the moon of no small importance to astronomy.

Some years ago Schmidt, at Athens, maintained that a volcano given out by Mädler as extinguished is no longer to be found, and he thereby proved the possibility of changes on the apparently rigid surface of the moon. If a photograph of the moon's surface could have been taken forty years ago, when Mädler observed the volcano, we should now be certain about this point, which is still hypothetical.

But the sun and its eclipses, the moon, and stars are not the only objects of astronomic photography. Its province extends further since the discovery of spectral analysis.

When it was discovered that those wonderful lines intersecting the sun's spectrum (see chap. vii. p. 63) were occasioned by glowing substances of different nature, and that each element shows invariably the same lines, so that the presence of certain spectral lines establishes the presence of certain substances, it became necessary to possess an exact view of the solar spectrum, with all its countless lines. This was essential, in order to be able at once to see what substances are yielded by these lines, by comparing this view with the spectrum of a flame, or of a fixed star. Kirchhoff, one of the discoverers of spectral analysis, and Angström have prepared such a detailed view of the solar spectrum. Their labour would have been materially simplified if Rutherford had published his photograph of the spectrum a year earlier.

I admit that this photographic spectrum of Rutherford's only shows the lines of the photographically operative part of the spectrum—from green to violet —but it does this with wonderful clearness. Many lines that appear faint to the naked eye show themselves strong and sharp in the view; nay, lines are discovered in the photograph of the spectrum which Kirchhoff did not see in the solar spectrum.

The causes of this phenomenon may be twofold: either the eye does not take in certain rays of light by certain lines,—as we know it is not influenced by the ultra-violet rays, which have a strong photographic effect,—or it is possible that changes take place in the sun, that at certain times fresh substances come to its surface, and thereby new lines become apparent.

The taking of a spectrum is effected with the help of

an ordinary spectral apparatus, seen in Fig. 79. This consists of a tube *A*, which has a fine slit *F*, through which the light penetrates. At the end of the pipe is a lens, which makes all the rays issuing from the slit parallel, and conducts them to the prism *P*. This refracts the rays in such wise that they fall into the tube *B*, and can be observed through its narrower end.

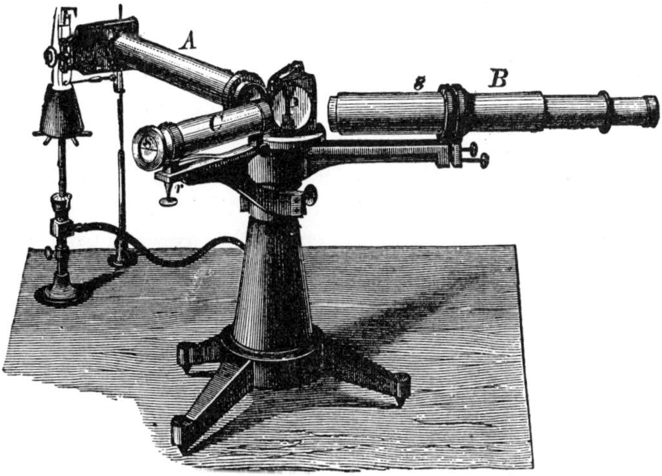

Fig. 79.

If the object is to photograph the spectrum thus seen, an opaque photographic camera is placed close to the tube, its eye-piece is drawn a little out, and then the image of the spectrum appears upon the ground-glass slide.

Attempts have been made to solve other important problems by the help of photography. Thus, Dr. Zencker hoped to be able to trace the path of falling stars by means of it. Unfortunately, these were found to give out too little light to produce, while they lasted, an impression on the photographic plate.

A grand new era is before photography in observing the transit of Venus.

In determining the distance of heavenly bodies, the earth's orbit is taken as base; therefore the knowledge of the exact amount of this base is assumed. Now this amount has only hitherto been determined by approximation, and is in round numbers one hundred and sixty millions of miles.

It has long been an effort to determine this number more accurately, but to do so is attended with great difficulty. Let it be conceived that there are at two different points of the earth, *a* and *b* (Fig 80), two observers who look with telescopes at a star *x*, and measure the angle which the eye-line makes with the line *a b*; it can be determined from both angles and the line *a b* (which is easily found) what the distance of the star is from *a* or *b*. This is the trigonometrical method, and it gives certain reliable results, if the distance of the star is not too great; thus, for example, the distance of the moon, which is about ten of the earth's diameter, is easily ascertained. If the star to be measured is too remote, the eye-lines *a x* and *b x* are nearly parallel, no difference exists between the two angles at *a* and *b*, and the trigonometrical method is useless. This is the case with the sun, which is ninety-five millions of miles from the earth. We can therefore only ascertain its distance by indirect methods.

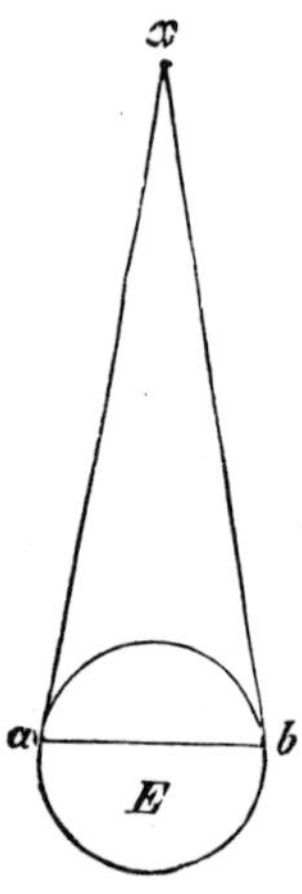

Fig. 80.

According to a law discovered by the celebrated astronomer Kepler, the squares of the periods of revolution of the planets are in proportion to the cubes of their distances from the sun. Thus, if the period of the earth's revolution is U, that of Venus u, the distance of the earth E, that of Venus e, according to this law

$$U^2 : u^2 = E^3 : e^3.$$

If the cube root is extracted from both we have—

$$\sqrt[3]{U^2} : \sqrt[3]{u^2} = E : e, \text{ hence,}$$

$$\sqrt[3]{U^2} - \sqrt[3]{u^2} : \sqrt[3]{u^2} = E - e : e.$$

But $E - e$ is the distance between the earth and Venus. When this has been determined by measurement, three terms of the proportion are known, for the periods of the revolutions of Venus and the earth are accurately known. Then, by the simple rule of three, the fourth term e can be found; that is, the distance of Venus from the sun. If to this is added the distance of the earth from Venus, we obtain the distance of the earth from the sun, which was required.

Thus the determination of the distance from the sun depends on that of the distance of Venus, which must be taken at the moment when Venus is between the earth and the sun. But Venus is not visible at the moment when it is placed before the sun's disc. This is only exceptionally the case—twice in every century—and then it appears as a small black point on the sun's disc, which, however, continually changes place, on account of the earth's movement and its own. This circumstance renders difficult the taking simultaneous measurements at two different and remote points of the earth, and therefore the idea has been entertained of using photography as an auxiliary. If by its help, and in the manner described above, a sun picture is taken during the transit of Venus, the distance of Venus from the sun's centre can be easily measured upon it. The centre of the sun is a fixed point that can be assumed to be stationary.

If the earth is supposed to be *E* (Fig. 81), Venus at *V*, and the sun at *S*, the observer at *a* will see Venus under the centre of the sun, underlying it, while an observer at

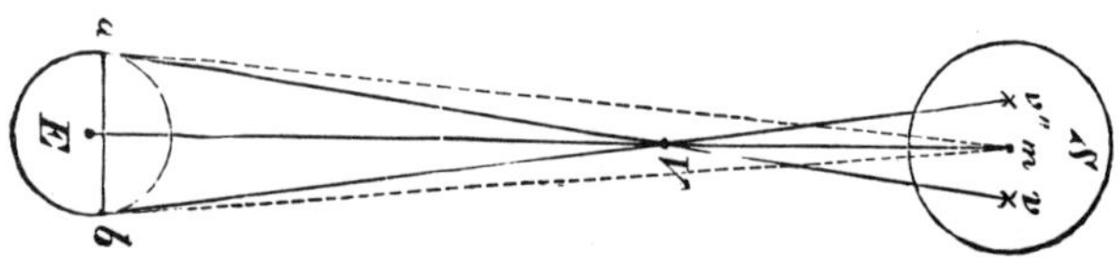

Fig. 81.

b will see Venus above it. Accordingly, Venus will present a different position to the sun's centre on photographs at various points of the earth.

Now, the situation of the line of inclination of the sun's centre is accurately known. The diameter of the sun corresponds to an angle of thirty minutes. If the sun's diameter is supposed to be divided into sixty parts, each part corresponds to the arc of a minute; therefore, it is only necessary to measure the number of such parts, separating Venus from the sun's centre, and we find directly the angle which the direction of the eye-line of Venus (for example, *a b*) makes with the direction of the eye-line of the sun's centre *a m*. If this angle is drawn from the angle which the eye-line of the sun's centre makes with *a b*, we obtain the angle which the eye-line of Venus makes with the line *a b*, and that gives all the data necessary to calculate the distance of Venus, and, from that, the distance of the central body which forms the foundation and base of all astronomical measurements.*

The determining of the angle by photography is of special value, as this measurement can easily be made, and at any convenient time, whereas measurements of the star itself can only be made while the phenomenon is visible, and hence many errors are introduced in the agitation of the moment. It is natural that measurements of this kind require apparatus of the most accurate description, and the adoption of many precautions; therefore preliminary experiments have now been begun to determine the degree of accuracy which a measurement by means of photography admits. If these preliminary experiments give a favourable result, numerous photographic expeditions will be sent out to observe the transit of Venus. Germany proposes to occupy five stations: Tschifu in China, Muscat on the Persian Gulf, Kerguelen's Land, and the Auckland Islands. Besides these, England, France, Russia, and America are equipping photographic expeditions which will occupy different points, and thus we may hope, though some stations are visited by unfavourable weather, still to obtain numerous plates by means of which the great astronomical question can be solved.

* Our space does not permit us to enter into the details of the method of determining the sun's parallax; it is only our purpose to give a plain intelligible statement of the principle of the thing. Those who are specially interested in the subject are referred to Dr. Schorr's "The Transit of Venus over the Sun." Brunswick: Vieweg. 1873.

Section V.

The Photographic Observation of Scientific Instruments.

Observations with the Thermometer and the Barometer—Neumeyer's Apparatus to determine the Depth of the Sea.

Meteorological observations require a daily repeated reading of the barometer and the thermometer. To economize this reading, and yet to receive a perfectly safe register of the state of the thermometer and barometer at each minute, photography has been turned to account. Let the reader imagine behind the tube of a thermometer *R* (Fig. 82) or barometer a drum, which revolves round its axis *a* by means of clockwork. Let sensitive paper be wrapped round this drum, and the whole, except the thermometer, be enclosed in a cylinder *S* which has only a small slit behind the thermometer, through which the light can penetrate. The upper part of the thermometer will let the light through, while the thread of quicksilver will stop the light. Therefore the strip of paper above the quicksilver will blacken, and the limit of the blackening on the paper will rise and fall with the mercury. Now the time can be marked beforehand on the paper. As the drum revolves once in twenty-four hours, the strip of paper need only be divided perpendicularly into twenty-four parts, and the first part be moved opposite the thermometer directly the clock strikes twelve, after which the whole may be allowed to revolve. Then the coloured strip shows

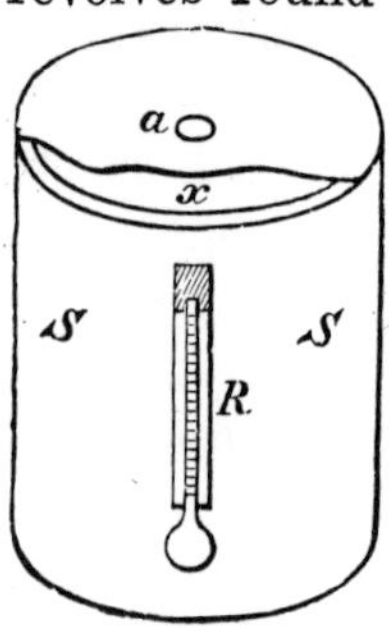

Fig. 82.

the height of the thermometer at all times of the day. In the same manner, the height of the barometer can be registered by photography.

Professor Neumeyer has latterly employed a similar instrument to determine the temperature in the depths of the sea. As there is no light producing chemical effects at those depths, Dr. Neumeyer sends down a light-producing apparatus. This consists of a galvanic battery, and a Giesler's tube; that is, a pipe in which very attenuated nitrogen gas is enclosed, and through which the electric current is passed. Then the tube gives out a faint light. But this faint light works chemically very powerfully, because it contains many of the invisible ultra-violet rays (see p. 64), and in three minutes it effects the blackening of the paper. Neumeyer also attempts to determine with his apparatus the direction of the oceanic currents. For this purpose the apparatus has an appendage not unlike a vane, and, as when suspended to a cable, it can be conveniently turned in all directions. If oceanic currents occur, it can be so placed in the depths that the vane is parallel to the current. A magnetic needle, within a tube impervious to water, is enclosed in the apparatus, and moves over a disc of sensitive paper; this magnetic needle points, of course, to the north, and the luminous tube above it marks exactly its position on the sensitive paper, which is firmly fastened to the box. Therefore, it can be easily seen what situation the apparatus has assumed with reference to the magnetic needle—that is, the north.

SECTION VI.—PHOTOGRAPHY WITH REFERENCE TO MEDICAL RESEARCH.

Photographs of the Interior of the Eye, the Ear, etc.—Stein's Heliopictor.

Photography has been begun to be applied on a large scale to the province of medical science, not only in taking interesting anatomical preparations and morbid phenomena of short duration, but in giving exact anatomical views of the different organs. The apparently impenetrable interior of living organs has been disclosed by eye-mirrors, ear-mirrors, and throat-mirrors, so that their interior is fully disclosed to the eye of the observer. In like manner, the image visible to the eye has been successfully retained by photography. Dr. Stein, of Frankfort-on-the-Maine, has done good service in this branch, not only as a practical photographer, but also by the construction of suitable apparatus. It would exceed the limits of this book to describe all the apparatus necessary for this purpose. We shall content ourselves with the description of one, that for taking the interior of the ear.

The apparatus consists of three parts: 1st, the ear-funnel *A*; 2nd, the lighting apparatus *B*; 3rd, the photographic apparatus *D*, with the lenses *C* (Fig. 83). These parts are placed together, as may be seen by the accompanying diagram. The apparatus is fastened by a joint to a corresponding stand, in order to give it the proper direction, according to the position of the sun. The ear-funnel *A* is a conical tube about 1½ inches in length, to push aside the small hairs which interrupt the view; it is made of vulcanized india-rubber. The lighting apparatus *B*, which is easily closed by a cover at *a d*, consists of two metal pipes, soldered together at a right angle at *b c*, of which one is provided with parallel sides, and the other with curved sides. At the place where the two tubes unite is a perforated aplanatic metal mirror (*e g f*), inclined at an angle of 45°.

The photographic apparatus *c* consists of a double objective *C* of twelve lines, besides a small camera, two inches deep. The ground-glass shade *X*, and the box *Y*, are adjusted in a rectangle *D*, easily moved. An enlarging plano-convex lens is situated between the objective and the lighting apparatus at *h*. According to the position of the sun, of a bright cloud, or any other

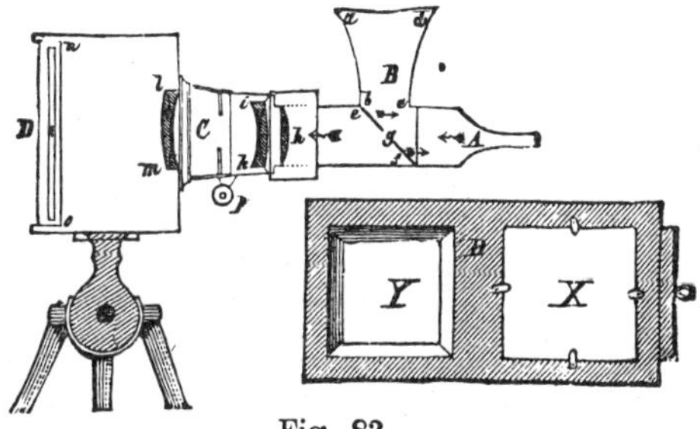

Fig. 83.

point of light, the lighting apparatus *B* can be moved by turning round on its axis; so that, in conjunction with the joint of the stand, the apparatus can be turned easily and firmly in all directions.

The rays which penetrate into the tube *B* are thrown by the perforated plane mirror *e f* in the direction of *A* on the drum of the ear. Reflected thence, they pass at *g* the perforated plane mirror, and the image of the drum of the ear is thrown on the ground-glass slide *n o*, by the combination of the lenses of the objective *h i k l m*.

The firm fixing takes place partly by means of the screw of the objective at *p*, partly by moving the lens at *h*, according as an enlarged image or one of life-size is desired. During the photographic process, an assistant must pull the ear muscle backwards and upwards, in order to give a proper direction to the funnel in the tortuous aperture of the ear. The time of exposure in the sunlight, if a good collodion of iodide of bromium is employed, lasts half a second; under bright clouds on a clear day, from five to ten seconds, according to the intensity of the light. The opening and shutting of the apparatus, to favour the operation of the rays of light, is effected at *a d*.

Dr. Stein has constructed, as an aid to naturalists and physicians, a very pretty instrument, called the heliopictor, which admits of taking views on moist plates without the dark chamber. The heliopictor is a kind of box which can be placed at the back of every camera. Dubroni, of Paris, first constructed such developing boxes. This box, a section of which is given in the diagram below, contains a glass receiver *K*, in which a silver solution can be poured through a stopcock, not visible in the figure. The glass plate to be prepared is covered with collodion, then brought through the door into the box placed at the aperture *O* of the glass case, and the door is shut. Then the spring *a* presses the water-tight plate *p* against the glass receiver. After this, the box is turned over to the right, the silver solution flows over the plate, and renders it

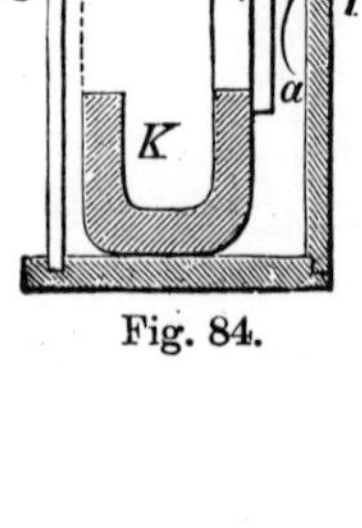

Fig. 84.

sensitive. The continuation of the operation is observed through a yellow glass slide *S*, which admits no chemical light. After the plate has been properly sensitized, the box is again placed upright, and brought into the camera instead of the ground-glass slide, *S* is drawn up, and light is admitted. Then the silver solution is drawn off through a stopcock, and a solution of green vitriol turned on instead; by tilting the box this flows over the plate and develops the picture. Its coming out is observed through the yellow slide *S*. After the developing, the picture is taken out and fixed.

Stein improved the developing box, by substituting a vulcanized india-rubber receiver, easily taken out and easy to clean, in the place of the glass receiver. He also introduced the method of filling and emptying the receiver by means of a stopcock, Dubroni having employed pipettes. Both apparatus are in the "Photographischen Mittheilungen," Jahrgang X. Nr. 117, 118.

Section VII.—Photography and the Microscope.

On Microscopes—Taking Microscopic Views—Their Application.

Nowhere has photography shown itself a more brilliant auxiliary or substitute for the art of drawing than in the reproduction of miscroscopic objects. This field was worked at the earliest period of art, for Wedgwood and Davy strove to retain the images of the solar microscope by the help of sensitive silver paper. This solar microscope seemed, in fact, to be made for photographic purposes. It consists principally of a microscopic object, which is inserted at *m*, and is either a drop of liquid brought upon a glass plate, or a small solid body compressed between two thin glass plates (Fig. 85).

The small lens projects an enlarged image of this minute object accurately on an opposite screen, or a white wall, exactly as shown in the following figure.

The screw at *D* serves to approach or remove the lens from the object *m*, and thereby to throw out the object sharply upon the screen. *E* is a dark window, by which the rim of the round image is cut off.

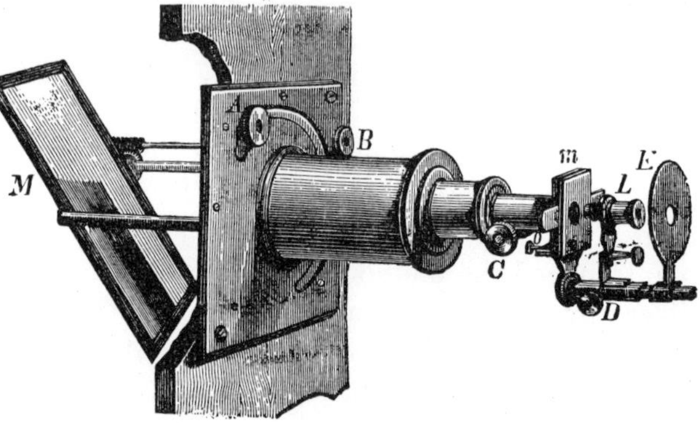

Fig. 85

The principal part of the image *B C* contains the lens that transmits the light. Each considerable enlargement diminishes the light of the picture materially; if it is enlarged three times, the clearness is diminished to $\frac{1}{9}$; if enlarged fourfold, to $\frac{1}{16}$; if fivefold, to $\frac{1}{25}$; if a hundredfold, to $\frac{1}{10000}$. With such a diminution of brightness the eye would not detect anything, if care were not given to throw an intense light on the object. The system of lenses contained in the pipe *B C* answers this purpose. This concentrates the sun's rays, which are reflected by mirror *M* into the tube *B*, on the microscopic object; and the latter becomes in this manner so well lighted that it admits of any amount of enlargement. The room in which the instrument is placed is dark; accordingly, all the conditions are present that enable proper photographs to be taken. All that is requisite is to place a sensitive plate instead of the image.

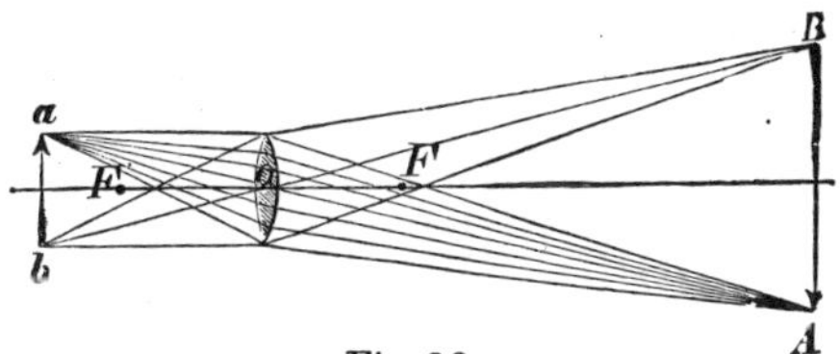

Fig. 86.

Few persons, however, possess a solar microscope. For ordinary investigations a microscope is used similar to that in the accompanying figure. It contains at *o* a system of enlarging lenses, which projects an enlarged image *S R* of the small object *r s*, as shown in Fig. 88. This is viewed through the eye-piece *c d* (Fig. 88), which is placed at *n* (Fig. 87), and enlarges for the second time the image *S R*, so that a still greater image *S′ R′* (Fig. 88) is produced.

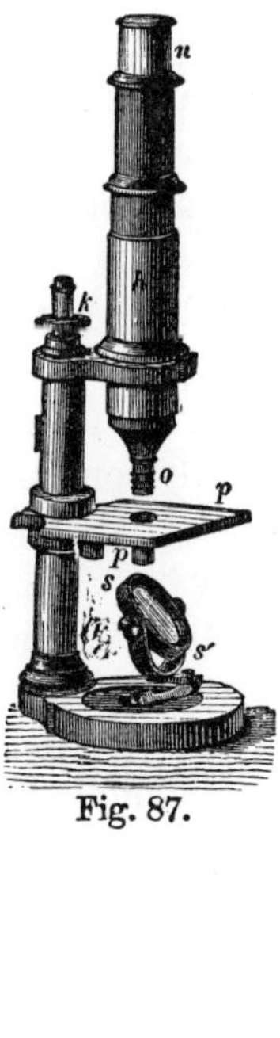

Fig. 87.

This is seen directly by the eye of tho observer. The necessary light is thrown on the object by the help of a concave mirror *s s′* (Fig. 87).

To produce photographs with the help of such a microscope, a photographic camera can be placed in a straight line with the eye-piece *n* (Fig. 87), by supporting it on a three-legged stand. This camera does not require a lens, like camera p. 90, but only a slit through which the opaque tube *n* passes. The eye-piece *n* (Fig. 87) is fixed into a sleeve-like appendage which surrounds the slit, and then the tube *h* is raised slightly.

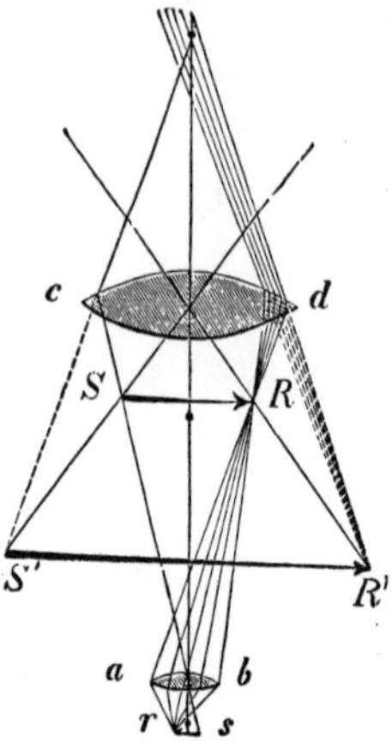

Fig. 88.

By this means the enlarged image of the object situated at *S R* (Fig. 88) is visible on the ground-glass slide of the camera, and can be easily photographed.

It is necessary in doing this that all light not emitted from the object should be excluded. If the mirror *s s'* (Fig. 87) throws light on the object, many rays pass beside it, fall on the lenses, and occasion reflections that materially disturb the purity of the image. In this case it is advantageous to insert a system of lenses between mirror *s s'* and the object, concentrating all the rays on the object.

Instead of sunlight, recourse is had to artificial light; for example, electric and magnesium light, which makes the observer independent of the weather. The beauty of the micro-photograph depends essentially on the beauty of the preparation to be photographed. This must be so arranged that it shows perfectly clearly all characteristic parts; all disturbing accessories, dust and so on, must be removed, for they are equally magnified with the object. A skilful preparer is therefore required to do anything good in micro-photography, which also depends on the excellence of the instrument, its proper arrangement, and the choice of the proper time. It is important in an instrument to correct the lenses for chemically operative rays. (See p. 190.)

Excellent results have been achieved in micro-photography by Neyt at Ghent, Girard and Lackerbauer in Paris, Fritzsch and Kellner at Berlin, and Woodward in America.*

Microscopic photography is of extraordinary use in anatomical preparations, which quickly change and become decomposed in chemical combinations. It is also useful in permanent bodies; thus, for the knowledge of microscopic crystals, which are enclosed in many kinds of stones, and show themselves clearly in thinly polished plates. In the images of these crystals, the angles can be easily measured with the help of a protractor, and from them the nature of the crystals may be inferred. Professor Gustavus Rose has reproduced in steel engravings many micro-photographs of this kind, taken by the author, in his treatise on meteorites.

Section VIII.

Microscopic Photographs and the Photographic Pigeon Post.

Nature of Microscopic Photographs—Their Importance for Libraries—Employment of the Pigeon Post.

Some years ago jewellery and toys were offered for sale in Paris, containing small magnifying glasses. If these were held before the eye, small transparent images, some of them portraits, and others writings, came into

* Full details are given in "Die Photographie, als Hülfsmittel mikroskopischer Forschung," by Moitessier and Bennecke. Brunswick: Vieweg.

view. These little images were what were called a microscopic photograph on glass. Such an image is by no means the representation of a microscopic object, but of a large-sized object, only the image is so small that a microscopic lens is required to view it. The production of these images does not differ from that of others; it only requires an instrument that projects images of microscopic minuteness in an optical manner, and this is effected by employing small lenses of very short focal distance. In using these a direct view of nature is not taken, but in the first place a photographic negative is prepared with the ordinary camera from the object chosen; after this, with the help of the small lenses, little microscopic images on glass are obtained with the ordinary collodion process. These little glass images are polished down, a small lens is fastened on them, and then they are set in metal. Such images are in themselves little else than toys, which can even be perverted, if, as has happened, immodest subjects, taken in this fashion, are given into the hands of unsuspecting persons—a fact that speedily brought this branch of photography into discredit. But there are circumstances in which such microscopic photographs can be of extraordinary value. Simpson in England has called attention to the fact that, by the help of photography, the contents of whole folios can be concentrated within a few square inches, and that the substance of books filling entire halls, when reduced by microscopic photography, can be brought within the compass of a single drawer—a circumstance which, with the enormous increase of material that has to be swallowed by our libraries, may be of importance. No doubt, to read such microscopic works requires either a microscope or an enlarging magic lantern.

Hitherto it has not been applied to this purpose, though Scamoni's heliographic process, described further on, would considerably facilitate the creation of such microscopic libraries. But such microscopic photographs have obtained great importance in promoting pigeon despatches. During the siege of Paris in 1870, the blockaded city held communication with the world outside by means of balloons and carrier pigeons. The first mode of communication was almost engrossed for political objects; the second only admitted the transmission of very minute writing. Letters, however condensed, could scarcely have been sent more than two or three at a time by a pigeon. In this case, microscopic photography presented a valuable means of concentrating many pages on a collodion film of only one square inch, and of expediting more than a dozen of such almost imponderable films packed in one quill. Dagrand at Paris, who first prepared microscopic photographs, also set going the system of these pigeon despatches. All the correspondence which had to be diminished was first set up in type, and printed together on a folio page. A microscopic photograph was made of this folio page, contained in about the space of 1½ square inches. This collodion film, with the image upon it, was then glazed over by pouring leather collodion over it; that is, collodion containing a solution of glycerine. This glucose collodion easily dries, separates from the picture, and forms a transparent film. A membrane of this kind could contain as many as fifteen hundred despatches. At the place of arrival these membranes were unrolled, and then enlarged by the help of a magic lantern; a number of writers thereupon set to work to copy the enlarged despatches, and ultimately forwarded them to their respective addresses. Thus Paris corresponded, by the aid of photography, for six months with the world without, and even poor persons were able to let their relatives know that they still lived.

Section IX.—Pyro-photography.

Fireproof Views—Their Production by Photography—Grüne's Method—Its Application for the Decoration of Glass and Porcelain.

An ordinary photograph is, as paper, very combustible, and exposed to injury from corrosive substances. Encaustic images on porcelain and glass do not participate in this exposure to injury, and therefore attempts have been made to prepare fireproof photographs, especially for the decoration of glass and porcelain. Success has crowned these efforts in several cases. One of the simplest processes in that of W. Grüne at Berlin.

Grüne found that the collodion image—which, as we have seen (p. 112) consists of minute parts of silver—is capable of manifold changes, and that, moreover, it is easily transferable, with its elastic collodion film, to other bodies. The film, with the picture, can be placed in different solutions, and then transferred to curved surfaces, etc. If the little collodion image is placed in a metal solution, a chemical change ensues. Assuming the metal solution to contain chloride of gold, then the chlorine passes over to the silver, of which the picture consists, chloride of silver is formed, and metallic gold is precipitated as a fine blue powder on the outline of the picture. Thus a gold picture is obtained.

With certain precautions this can be transferred to and made encaustic on porcelain. By this means an unpolished image is obtained, which can be rendered brilliant by polishing. Grüne has employed this to produce gold ornaments on glass and porcelain. Drawings and patterns of different kinds are photographed; the image obtained is changed into one of gold, then burnt in, and thus the most beautiful and complicated decorations can be produced without the assistance of the porcelain painter.

If a silver picture be plunged into a solution of platinum instead of a solution of gold, a platinum image is obtained. This assumes a black colour on being burnt into the porcelain. In this manner black portraits, landscapes, etc., have been produced on porcelain.

Images of this kind in other tints can be represented as black. For example, if the image is dipped in a combined solution of gold and platinum, the gold and platinum are precipitated on the picture. The image thus obtained, if burnt in, presents a very agreeable violet tint.

Solutions of uranium, of iron, and of manganese effect precipitates on a collodion picture, modifying its colour, and, when burnt in, producing different brownish or blackish tints. We shall see, later on, that there are other means of producing such pyro-photographs. Details will be found in the chapter on the photochemistry of chromic combinations.

Section X.—Magic Photography.

Invisible Photographs—Magic Pictures and Magic Cigar Ends.

Closely connected with Grüne's process for producing porcelain pictures is what is called magic photography. A few years ago small white sheets of paper were offered for sale which, on being covered with blotting-paper and sprinkled with water, displayed an image as if by magic. The white sheets of paper, to all appearance a blank, were photographs which had been bleached by plunging in chloride of mercury. If a photograph not containing gold—all the usual paper photographs contain gold—be plunged in a solution of chloride of mercury, a part of the chlorine passes over to the silver of the picture, and changes this brown mass into white chloride of silver, which is invisible on the white paper. At the same time a chloride of mercury containing less chlorine,—hypo-chloride of mercury,—which is also white, and therefore invisible on the white paper, is precipitated. Now, there are different substances which colour black this white hypo-chloride of mercury. Among these are hypo-sulphite of soda and ammonia. If, therefore, the invisible picture is moistened with one of these substances, it is coloured black and becomes visible. In the magic photographs formerly sold there was hypo-sulphite of sodium in the blotting paper; this became dissolved on moistening the paper, or water penetrated to the magic image lying under it and made it visible.

Quite a different kind of magic photograph was offered for sale some years later—the magic cigar tips. These contained a small sheet of paper between the cigar and the mouthpiece, which the cigar smoke penetrates; with continued smoking an image became visible on the sheet of paper, which contained a magic photograph of the kind described above. The image was brought out by the vapour of ammonia which is in cigar smoke, and which has also the property of colouring black the magic photographs.

The magic photographs of recent times were introduced at Berlin by Grüne, but their principle was known before, as J. Herschel had produced some in 1840.

Section XI.—Scamoni's Heliographic Process.

Defects of the Positive Silver Process—Advantages of the Typographic Press—Relief of the Photographic Negative—Impress of the Same on Copper.

It was stated at an earlier page, that the positive photographic process had the defect of working very slowly. Every picture that has to be copied from a negative must have a longer or shorter exposure to the light. Hence, the worse the light, the longer the time required. This is of no moment with a dozen portraits, but if hundreds or thousands are to be prepared, time is of consequence.

Another disadvantage in the silver copy is its high price and doubtful durability. Attempts have been made, since the discovery of photography, to overcome these defects by combining it with printing press processes—lithography or metal-type press. To carry out the metal press, an engraved plate is used; that is, a metal plate in which the drawing is deeply incised. This is covered with engraver's ink, the ink penetrates into the incisions, and under a heavy press a steel engraving or copper-plate engraving is thus produced. Impressions of this kind can be made in a short time in great quantities, without the help of light, and without employing expensive salts of silver. We showed in the first chapter (p. 10) that an incised drawing on a metal plate can be made with the help of photography. We mentioned asphaltum as an auxiliary to this end. But the same object can be attained in another way; and one of the most original is that of Herr G. Scamoni, at St. Petersburg, the able heliographer of the Imperial Russian expedition for procuring State papers.

He observed that an ordinary photographic negative does not form a plane surface, but appears in relief; the transparent places—shadows—being in basso, and the light in alto relievo. But this relief is very faint. Scamoni tried to increase it by treating the freshly taken and developed picture with pyrogallic acid and solution of silver. In this manner fresh silver powder was precipitated on the picture, which has the property of attracting and retaining silver separated chemically. The relief was considerably increased by this strengthening process. It can be augmented by a treatment with chloride of mercury and iodide of potassium, which conduct the metal silver of the picture into more solid combinations. A relief was ultimately obtained nearly as high as that of the incisions of an engraved copper-plate. When a linear drawing has been taken in this way, and after the negative has been obtained, a positive has been prepared by repeating the collodion process in the camera, and the latter has been brought out enough in high relief by strengthening, all the means are in hand to produce an engraved copper-plate from the image thus received. The relief-like photographic plate is brought into a galvano-plastic apparatus, of which we shall speak farther on. This produces on the plate a connected copper precipitate, which is in basso or low relief where the plate shows high relief; that is, where there are strokes or outlines. Thus a copper-plate is obtained from which impressions can be taken as well as from one that has been engraved. This process is now used to reproduce drawings like copper plate.

Maps are prepared in this manner, in which the drawing can be photographically enlarged or diminished; also writings on an enlarged and diminished scale. Scamoni has thus reduced all the manifestoes of the Emperor Alexander, as also a page of the illustrated journal "Ueber Land und Meer" on leaves of one inch in width, on which the writing is perfectly legible through the microscope. Helps of this kind are not mere play, but they have a great importance for the preparation of paper money and for libraries, as we showed at pp. 11 and 210.

Section XII.—Photography and Jurisprudence.

Photographic Authentication Cards—Photographs of Criminals, of Railway Accidents, Fires, Documents, etc.

The application of photography to jurisprudence is of great interest. The faithful likeness of a man, or of an object, makes their recognition more certain than the most circumstantial description in words; and photography gives us such. Accordingly, repeated attempts have been made to utilize it as a means of authentication. This was first attempted in 1865, when the season tickets for the photographic exhibition at Berlin contained the portrait of the holder, that they might not be transferred. This plan is now adopted in the season tickets of the Zoological Gardens of Berlin. It is still more important for the recognition of criminals. Persons subject to various penalties are now photographed in prisons, partly as a means of recapture in case of evasion, partly to detect them in case they should be again brought in under a false name.

Justizrath Odebrecht, in a treatise on jurisprudence, recommends the taking of the bodies found, and in case of murder that of the victim and the surroundings, for the information of judges. This has been repeatedly done. Further, railway trains that have suffered an accident, buildings that have been destroyed by fire or the elements, are photographed for the information of railway and insurance companies, or of the legal authorities. Photography is very advantageous in this matter, through the rapidity of its operation, which can be completed within a few minutes, and carried on even during the restoration of the building. It is also of value in jurisprudence, by detecting forgeries. Very frequently forged cheques are photographed in order to send a copy for the information of those interested. Stolen and recovered articles are also often photographed to bring them to the notice of the proprietor. In many large cities the police cause pickpockets and sharpers to be photographed, and show an album of this kind to persons who have been robbed.

SECTION XIII.—PHOTOGRAPHY, INDUSTRY, AND ART.

Photography as a Means of Artistic Culture—Extension of the Art of Drawing through Photography—Pattern Cards—Application of it to Building Plans—Estimation of Solids by Photography.

We have already laid stress upon the importance of photography in works of art. It makes every work of art accessible to persons of slender means, and therefore it has become as important an auxiliary for popular culture in the province of art as the printing press is for science.

Photography is equally important in those branches of industry in which graphic representations are indispensable; for example, architecture and the construction of machinery. In their case photography forms an enlargement of the art of drawing, effecting in a few minutes what the draughtsman could only accomplish in several hours or days, and representing with a faithfulness to which no draughtsman could attain. In this connection we have already described, in our second chapter, the technical importance of the *licht-paus* process. It is the easiest kind of photography, but it only gives copies of the size of the original; however, the negative process allows an enlarged or diminished copy to be taken of every drawing, according to option. Photography is already very generally used for these reproductions. It is equally important for taking views direct from nature, be they machines or parts of machines, buildings or parts of buildings. Pictures of this kind present not only a graphic image, but they serve for instruction and demonstration in lectures. Nay, when a house is to be photographed, and measures have been laid down for its length, breadth, and depth, the dimensions of the particular part can be taken from the photograph by making allowance for the foreshortenings in perspective. Pictures on a small scale are commonly sent out as specimen cards. Iron foundries, manufactories of bronzes and of porcelain frequently issue lists of prices with photographic illustrations, of which the images are multiplied from negatives of originals by the printing of light. (See the following chapter.)

Further, the original application of photography is that relating to the plans of buildings. Architects who are at a distance from a building under their direction cause photographs to be taken every week, giving them a clear picture of the progress of the building. We have already hinted at the services that photography can render in the manufacture of porcelain, and further in combination with the multiplying arts. We shall learn more on this subject in the following chapter.

CHAPTER XV.

CHROMO-PHOTOGRAPHY.

WE have given a full account, in the first part of our book, of the chemical and physical principles of photography with salts of silver, and of its application to art, science, life, and industry.

Numerous attempts have been made to substitute other sensitive materials for the expensive salts of silver, and some of these attempts have been crowned with success. It is true that no substance has been hitherto found permitting a negative to be prepared in the camera as easily as iodide of silver. For the production of camera pictures from nature we are exclusively confined to iodide of silver and bromide of silver. But the case is different with the production of positives from negatives already existing. These can be successfully produced, not only by the help of salts of silver, but also of other metallic combinations. I admit that the results obtained are inferior in beauty to the silver pictures, but we shall see, later on, that they admit of multiplication through combination with printing by impression without the help of light. We shall now describe the most important of these processes.

Section I.—Chromic Combinations.

Oxides—Combinations of Chromium with Oxygen—Oxide Salts of Chromium—Protoxide and other oxides of Chromium—Chromic Acids —Chromic Acid Salts in the Light—Ponton's Discoveries.

A black mineral called chrome iron ore occurs in nature, especially in Sweden and America. If this be dissolved with carbonate of potash, a beautiful orange-red salt is formed, which dissolves in water and easily crystallizes on evaporation. This orange-red salt is the bichromate of potash. It consists, as implied by the name, of chromic acid and potassium. The latter is the chief component part of our potash; the former consists of a metal, like iron, and of oxygen. Chromium and oxygen form together very different combinations.

28	parts	chromium	with 8	parts	oxygen	to	protoxide of chromium.
28	„	„	„ 12	„	„	„	sesquioxide of chromium.
28	„	„	„ 16	„	„	„	suboxide of chromium.
28	„	„	„ 24	„	„	„	chromic acid.

The last combination, chromic acid, is the best known of all; on adding to it sulphuric acid, it changes to chromate of potash, and crystallizes into red needles, which easily lose part of its oxygen. For example, if chromic acid is dropped upon alcohol, the latter becomes inflamed, because it immediately withdraws oxygen from the chromic acid and changes it into a green body, the oxide of chromium. The oxide of chromium forms salts with acids; for example, sulphate of chromium. This unites again readily with sulphate of potash to form a dry salt, which is known by the name of chrome alum, and is sold crystallized in very beautiful dark violet octahedra. It is employed in painting and dyeing, together with chromate of potash.

If chromate of potash be mixed with a solution of green vitriol, the green vitriol takes up a part of the oxygen of the chromate of potash, and a brown protoxide of chromium is precipitated. This is often formed by the operation of substances absorbing oxygen on chromic acid or its salts.

Chromic acid is of special interest in the object that engages it, because both it and its salts are sensitive to light. Pure chromic acid, and also chromate of potash, do not change in the light; they can be exposed for years to the sunlight without any decomposition being perceived. As soon as a body is present that can be united with oxygen—for example, wood-fibre, paper, etc.—the light immediately produces its effect. This fact was published in the year of the discovery of photography, 1839, by Mungo Ponton, and in the "New Philosophical Journal" he writes :—

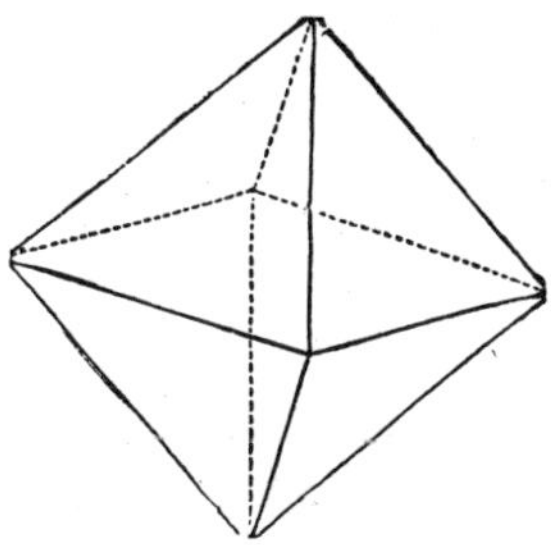

Fig. 89.

"If paper is saturated with a solution of chromate of potash, it becomes sensitive to the sun's rays. If an object be placed upon it, the part exposed to the light quickly assumes a yellowish-brown tint, shading more or less into orange, according to the strength of the light. The part covered by the object retains its original clear yellow colour, and the object is imprinted as a clear outline on a dark ground, with different shades of colour, varying with the different degrees of transparency of different parts of the object. In this condition the picture, though very beautiful, is not lasting; to fix it, it is sufficient to plunge it in water, whereupon all parts of the salt that were not touched by the light are quickly dissolved, while those on which the light could operate are perfectly fixed on the paper. By the last process a white picture is obtained on an orange ground, and it is perfectly durable. If it be exposed for many hours, the ground tint loses intensity of colour, but not more than happens with other coloured matters."

It appears that Mungo Ponton made experiments, like Talbot, in the first period of silver photography. Perhaps he also copied leaves (see p. 5). The copies which are produced in the above manner on chromate of potash are, however, immeasurably fainter than the copies on silvered paper.

They can at any time be easily prepared, by plunging a piece of white paper into a solution of chromate of potassium in the dark, by the light of a lamp. After a minute, they are taken out and suffered to dry. It is best to suspend them, and the dried surface is exposed to the light in a copper frame under dried leaves, or a drawing, or a negative.

The chromic acid is then reduced to brown protoxide of chromium, but if the exposure lasts very long the reducing process goes further, and a green oxide of chromium is formed. In this case the picture appears fainter.

Accordingly, Ponton's experiment remained a mere curiosity until the inventor of photography on silvered paper discovered another property of chromate of potash, which led to the most extensive applications.

This property consists in the operation of the combinations of chromium on glue.

SECTION II.—HELIOGRAPHY WITH SALTS OF CHROMIUM.

Properties of Gelatine—Chromate of Potassium and Glue—Talbot's Discovery—Effect of Light on the Solubility of Glue—Photographic Steel Engraving—Pretsch's Photogalvanography—Printing in High and Low Relief—Importance of the Former—Difficulty of producing Half-Tones by the Heliographic Method.

Glue in its purest form, known by the name of gelatine, is insoluble in cold water, but it sucks up cold water like a sponge, and thereby swells. If it is warmed with water, it dissolves; but on cooling the solution hardens to a jelly. This property is used to thicken soups. If to the warmed solution of glue be added aluminum, or a salt of the oxide of chromium, or chrome alum, the glue becomes insoluble in water, and forms a precipitate. On this is based the well-known system of white tanning; for in the tanning of a piece of leather the aluminum combines with the gelatine contained in the leather,—chondrin,—and this becomes thereby insoluble, and at the same time durable.

Chromate of potash and glue can be dissolved together in warm water in the dark, without the glue suffering from the chromic salt. If a plate or a sheet of paper be covered with a solution of chromate of potash and the film be allowed to dry, it becomes firm, and yet remains soluble in water as long as it is kept in the dark. But as soon as the film is influenced by the light, the chromate of potash is reduced to oxide of chromium, and this tans the film of gelatine; that is, makes it insoluble in water.

This observation was made by Fox Talbot in 1852, and, as a careful observer, he knew directly how to turn it to account. He coated a steel plate with a solution of chromium and gelatine, let it dry in the dark, and then exposed it under a drawing or a positive glass picture. The black lines kept back the light. Accordingly, at these places the gelatine remained soluble, but it became insoluble at the white places, through the operation of light. After the exposure he washed the plate in the dark with warm water. By this means the places that had remained soluble under the black lines became dissolved; the others were retained on the plate. Thus Talbot obtained a drawing on the metal itself on a brown ground. This is worthless by itself, but it provides the means of producing a steel plate for engraving.

We have already explained, at p. 215, the nature of steel engraving and copper-plate engraving. Both processes consist in the production of a metal plate which contains, in incised lines, the drawing that is to be reproduced. These lines become coloured when engravers are used, and imprint it upon the paper. The hard steel plates have the advantage of lasting for many more copies than the softer copper plate; only the steel engravings are far inferior to copper-plate in artistic beauty, and therefore the former have lost favour. But the steel engraving is very important to prepare technical and scientific diagrams, paper money, and the like, as less artistic beauty is required in their case. It was steel plates of this kind that Talbot produced by the help of light.

We have seen that his steel plate was covered with an insoluble film of gelatine, and that the metal was uncovered at all places where the light had not operated. He poured on it a fluid which ate into the steel; for example, a mixture of acetic acid and nitric acid. This mixture, of course, only took effect where the steel was exposed, and thus produced an incised drawing in the steel plate, so that the latter, after being cleaned, gives as good an engraving as if it were the work of the engraver.

Thus a new process was found to replace the difficult work of the copper-plate engraver by the chemical operation of light.

We have mentioned in the first chapter a similar process, based on the application of asphaltum; also a different one by Scamoni. (See p. 215.)

This discovery of Talbot was soon followed by a more productive one on the same ground.

An Austrian, Paul Pretsch, prepared, in 1854, copper plates by a similar process, with the help of galvano-plastic. He also took a film of gelatine, which contained chromate of potash, exposed this under a negative or a positive picture, and then washed it in hot water.

After doing this all the places were retained which had become insoluble through the light, and after the washing and drying they stood out in high relief.

Accordingly, in copying under a positive, the lines which were black in the original appeared in low relief, and the white parts in high relief.

This kind of film in relief was placed in a galvano-plastic apparatus. This apparatus has the property of precipitating copper or other metals on a surface. It consists of a galvanic element, as described p. 71, with whose pole a trough with a solution of copper and vitriol is united. To the zinc pole are suspended, by means of the rod *B* (Fig. 90), the reliefs which it is desired to imprint, after they have been made conductors by a coating of graphite; a copper plate is suspended at the copper end *D*. As soon as the galvanic stream operates, the fluid is decomposed. The copper adheres to the relief, and the thickness of the copper depends on the time the current is allowed to last. Accordingly, plates of any thickness can be produced.

If the original form was in low relief, the galvanoplastic impression will be in high relief, and *vice versâ*. Therefore, in the above case an impression is received with lines in high relief.

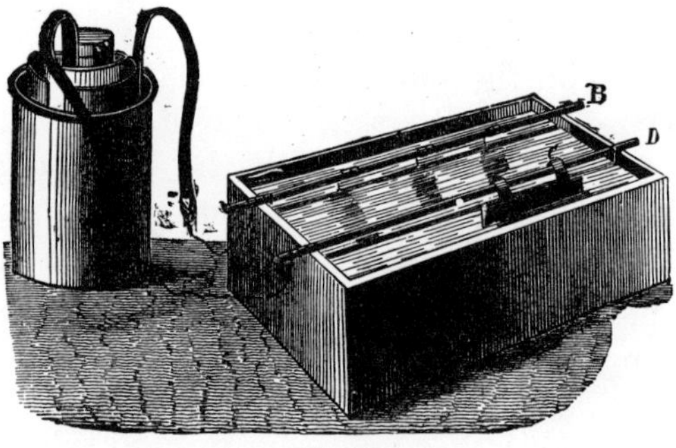

Fig. 90.

This kind of plate is also adapted to give impressions, but rather differently from an incised copper plate.

In an incised copper plate, the engraver's ink is rubbed into the incised marks, and then under strong pressure conveyed to paper.

In a plate with a drawing in high relief, the impression takes place as in printing; the raised places are rubbed over with printer's ink by the help of a leather ball, or of a cylinder blackened with ink, and then imprinted on paper. Letter-press is produced in this manner; all its letters are in high relief, also all woodcuts which accompany the text.

The printing-press is the simplest and cheapest mode of multiplying copies. It admits of the use of cheap papers, whilst copper-plate engraving requires a thick, soft, special paper. The printing-press, moreover, admits of woodcuts being printed in the text, whilst copper-plate printing requires special tables. Lastly, the printing-press works with extreme rapidity—steam press—whereas copper-plate printing requires much more time.

Further, the printing-press does not use up the type rapidly, as it works under feeble pressure; while copper-plate printing, which requires strong pressure, wears the plate considerably, so that after striking off a thousand copies, the impressions are no longer as good as at first.

The production of plates is very important for the printing-press, and Pretsch has taken the lead here. His process did not, indeed, produce the most perfect results. The low-relief plate which he produced on the gelatine film by the help of light was not deep enough to produce a high relief with galvanic impression; but this is necessary, for otherwise the printer's ink penetrates into the incised parts, which ought to remain white, while the washing of the proposed chromo-glucose pictures with hot water easily dissolves the finer particles of the picture, and this detracts materially from the value of the copies. Moreover, the printing with the help of galvano-plastic has its difficulties. The film of gelatine liquifies in part and loses its form. In short, the affair is not so simple as it appears; little difficulties exist, and these occasion errors which the unprofessional hardly observe, but which considerably diminish the effectiveness of the picture.

At an early period it was found that these processes offered a special difficulty, viz. the reproduction of the transitions from light to shade—the half tones. These were so very imperfect that the representation of natural objects—portraits and landscapes—was speedily given up, and people confined themselves to representing drawings, maps, and the like, on an enlarged or diminished scale, and thereby to producing stereotype plates for copper engraving and printing. This application is of no little importance, for it prepares a metal plate for printing, by the help of light, in as many hours as an engraver requires days, and at far less cost.

We add two plates to the present work, which by the help of gelatine and salts of chromium, by a modification of the process now described, have been carried into effect by Scamoni at St. Petersburg. Both are impressions of heliographic plates: the smaller one—Plate III., "Am Rhein"—a plate in high relief, which is printed in the letter-press; the other—Plate IV., "Johannisfest"—a low relief, in the style of copper engraving.

The Section III.—The Production of Photo-reliefs.

Photo-sculptures—The Pantograph—The Fount Process—Chromogelatine-relief—Fount-relief by Cold Water—Relief by Cold Water—Difficulty of its Production—The Transfer Process.

More than ten years ago intelligence was received from Paris of an entirely new discovery—photo-sculpture—which was said to produce statues by the help of light.

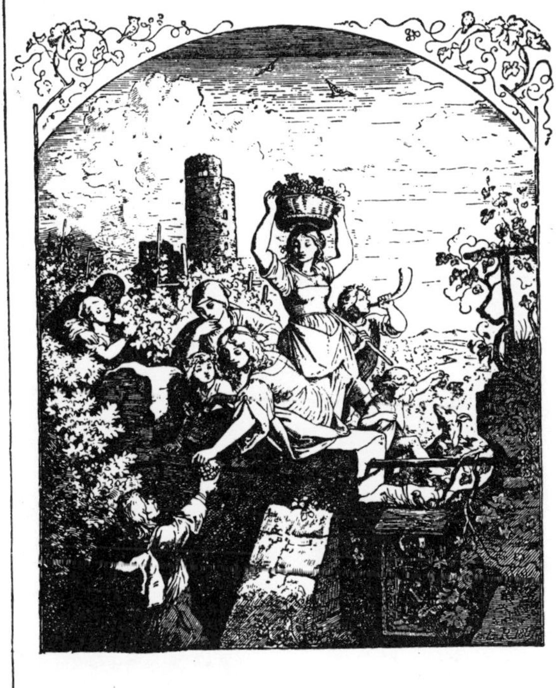

Johannisfest.

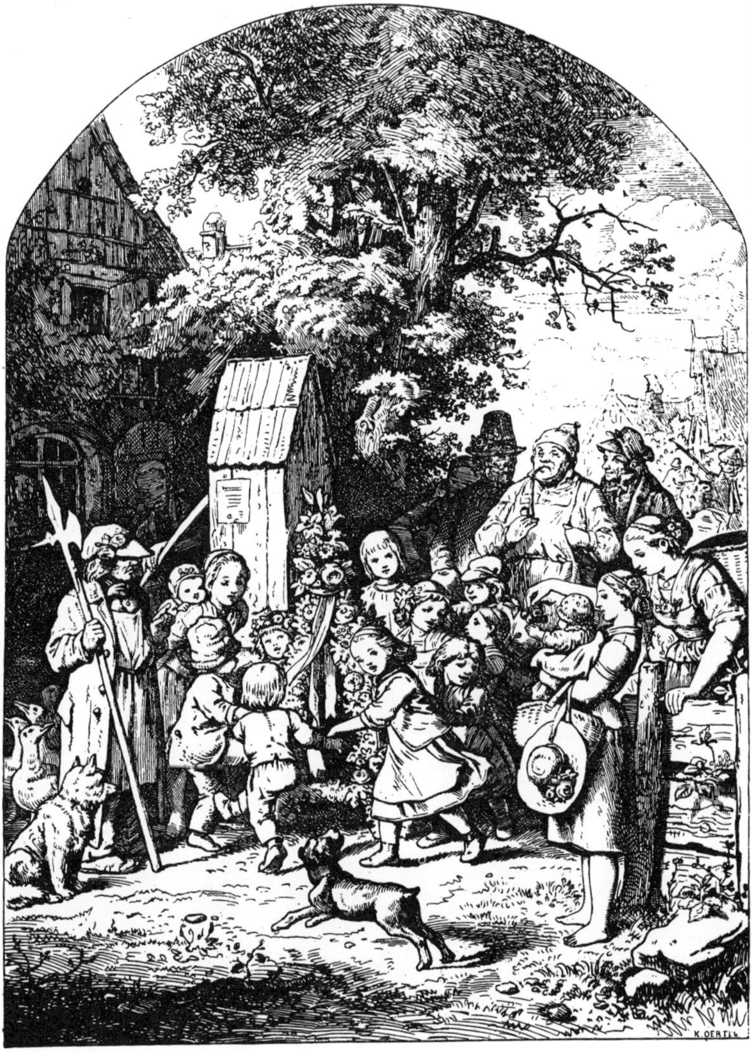

According to the description, this was effected by a circuitous process: a person was placed in the middle of a circle, and around him were placed about twenty photographic apparatus, which at a given moment took twenty pictures of the person, and represented him on every side. These photographs were afterwards transferred with their outlines to clay, by means of an instrument commonly called a pantograph. This consists of a system of bars *a b c d* (Fig. 91). Of this system one bar is placed at a fixed point *x*, the others are movable at the joints; *m n* are two pegs. If one peg *m* is carried

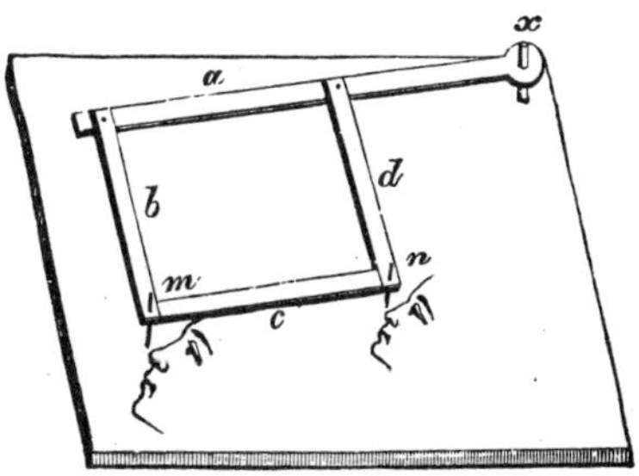

Fig. 91.

along a drawing, the other peg *n* makes the same movement, and, if a piece of paper be placed under the peg *n*, draws exactly the same line which the first peg *m* describes. If instead of the peg *n* we conceive a knife, which cuts out in clay the outline described by the first peg *m*, a profile is obtained in clay by moving the peg *m* along the circumference of an image, and in this manner all the outlines of the person taken can be transferred to clay. This photo-sculpture, as it is called, can only be carried out imperfectly. A careful manipulation by a very clever artistic hand is necessary for the work, and is indeed the essential matter. As far as the author has examined the matter, the pantograph is a mere pretence. A clever artist models the bust according to the photograph at hand.

Nevertheless, there are reliefs produced by light, and these reliefs are not inventions of advertisers; they are easily produced, and it is surprising that the process has not yet made a stand.

We have explained above the properties of gelatine, and remarked that it has the capacity of liquifying in cold water. This property is lost if the gelatine is saturated with chromate of potash, and exposed to the light. If this exposure is made under a negative, all the places situated under the transparent parts lose their capacity of liquifying, while the other places not affected by the light retain it. Accordingly, if the exposed film be thrown into water, the places which are not affected by light swell, whilst those affected by light remain in low relief. The result is a true relief,—the lights are in high relief, the shadows are in low relief, and this is so strong that it can be cast in gypsum. To this end the relief is dried with blotting-paper, rubbed with oil, and then a paste of gypsum is poured over it. This soon hardens, and gives an impression of the gelatine relief, being in high relief where the gelatine relief is so, and the contrary where it is in low relief.

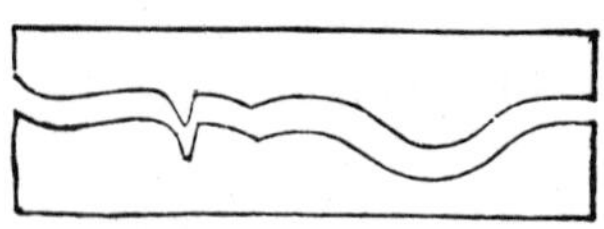

Fig. 92.

It appears as if a printing plate might be easily obtained for letter-press from such a gelatine relief. Let the gelatine film under a drawing be supposed to be exposed. The black lines then keep back the light; accordingly, the gelatine particles come out in high relief on being wetted with water. The drawing is therefore represented in high relief, and this is exactly what the printer requires; nothing further would now be required than to recast the relief in gypsum, and recast the gypsum form in metal, as happens daily in the stereotyping of woodcuts. But unfortunately this process breaks down, owing to a trifling circumstance,—the strokes are of unequal length in the relief wetted with water. But the letter-press requires the strokes to be on a plane surface, otherwise they cannot be equally inked and printed.

On the other hand, the casting can be very well applied as a picture in relief, if suitable *retouches* are given to it. Reliefs of this kind with portraits were sold some years ago as sealing-wax, but the execution was very imperfect, therefore these reliefs soon lost favour. Metallic forms are prepared according to this process in the Imperial Russian expedition for the production of State papers, and these forms can produce, when printed on paper, water-marks which are used in the production of bank-notes.

But reliefs can be obtained in another way, from an exposed film of gelatine, mainly by hot water. As we have seen above, this dissolves the parts which, not having been affected by light, have remained soluble, and it leaves the parts affected by light, and therefore insoluble. These parts that have remained insoluble stand out as prominences.

Another precaution is necessary. Suppose that *N* (Fig. 93 *a*) is a negative, that *c c* are its transparent parts, and *b* the semi-transparent, what are called half-tones. If a film of gelatine and salts of chromium *g* (Fig. 93 *b*) is exposed under them, the light penetrates in various degrees, according to its strength—most in the transparent places, less in the half-transparent, and not at all in the opaque parts.

Accordingly, insoluble films of different thickness will

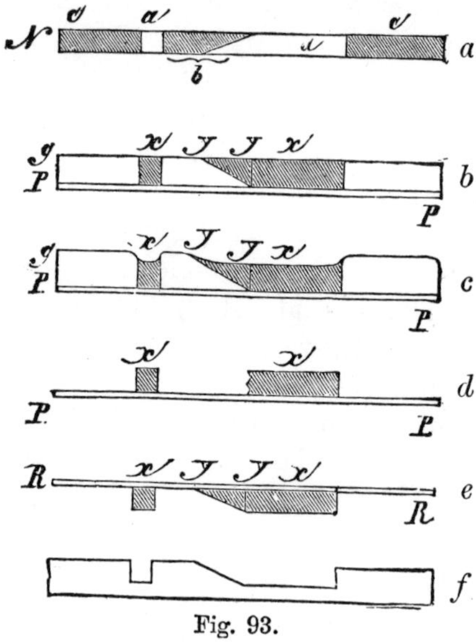

Fig. 93.

be formed, as represented Fig. 93. The shaded parts in the figure denote the portions that have become insoluble.

If now the film of gelatine (Fig. 93 *b*) is plunged in hot water, all the parts left white in the figure become dissolved; but at the same time the half-tones not adhering to the substratum *P*—for instance, paper—become detached and are torn off. Therefore a relief of the form Fig. 93 *d* remains behind; the half-tones are wanting (*y y*). In order to avoid this interruption the new substratum must be given to the exposed surface, which retains the half-tones. For this purpose a piece of albuminized paper is pressed on the exposed surface, after being very firmly fixed to the upper surface of the gelatine film. If the sheet (Fig. 93 *b*) is plunged in hot water, the membrane *P* becomes detached from *g*, the little portions of gelatine remain suspended to the albuminized paper, the white places in Fig. 93 *b* become dissolved, and all the half-tones *y y* adhere firmly to the new layer, as in Fig. 93 *e*, and form a relief. To economize these, the gelatine film can be produced on a transparent collodion film, and exposed to the light on the reverse side under the negative; the result is then the same as in Fig. 93. This process is named the transferring process. If the relief produced by cold water sprinkling, described p. 232 (Fig. 93 *c*), is compared with that produced with hot water (Fig. 93 *e*), the difference is at once apparent: in the former case the parts not exposed stand out in relief, in the latter case those exposed to light.

Section IV.—Printing in Relief.

Production of Photographic Half-tones—Production of a Printing Plate in Relief from a Gelatine Relief—Woodbury's Printing Process—Its Importance—Printing in Relief on Glass, and Magic-lantern Pictures.

The production of reliefs with cold and also with warm water, which we have described in the previous chapter, did not lead the way to a kind of photo-sculpture, but to a peculiar process of printing, which has been called printing in relief, from its mode in operation, and was invented by Woodbury in England, in 1865.

The heliographic methods of printing appear to be very simple, but they are unable to reproduce images of all objects on plates that can be used for printing. An outline drawing or letter-press can be tolerably well reproduced by this method, on an enlarged as well as a diminished scale, and it is this that gives value to the process. But it is much more difficult to reproduce in this manner, from nature, pictures with half-tones; for example, stippled drawings and photographs. The tender half-tones become rough and hard, rendering the picture very ugly. According to Osborne, the cause of this is found chiefly in the nature of half-tones in copper-plate printing. The half-tone in copper-plate printing is produced by the fact that black strokes of various thickness are placed beside each other, as can be detected in engraved copper-plates and in ordinary woodcuts; or it is effected by roughing the plate, thus forming a series of points, which appear more or less grey or black, according as they are nearer to or farther from each other, and thus they form the half-tones. The half-tone of stippled drawings and photographs is quite different. It does not form strokes or points, but a homogeneous light or dark colour.

Accordingly, it was first necessary in a manner to break up the photographic half-tone into a series of strokes or points, in order for it to become a copper-plate half-tone, and this constitutes the difficulty.

Woodbury conceived the idea to produce, by a new

printing process, homogeneous half-tones, perfectly similar to those of photographs or stippled drawings.

He represented a relief, by exposing a film of chromo-gelatine resting on collodion under a negative, and on the reverse side, and by treating it with hot water. (See last chapter.) This relief shows the inky parts of the original in high relief, and the light parts in low relief. For the negative is transparent where the original is black; hence the light passes unimpeded through those places. The half-tones imprint themselves by varying between high and low relief. They are, as it were, the declivities of the heights. (Compare Fig. 93 *e*).

If this gelatine relief is suffered to dry, it becomes wonderfully hard and firm. It can then be placed with a plate of lead under a strong press, and an impression of the relief can be thus obtained in lead. The prominent parts of the gelatine relief appear, of course, depressed in the lead, and the depressions prominent, as represented Fig. 93 *d*.

Woodbury uses this lead relief as a printing plate. But he does not print it off with oily printer's ink, for it is too opaque, but with a semi-transparent gelatine ink. This is poured warm on the plate in a horizontal position, it penetrates into the depressions, and now, if a piece of paper be placed upon it and pressed gently down, the gelatine consolidates quickly, and an impression resembling relief is obtained on the paper. As the ink is transparent, it appears in its thin coats much less black than in the thick coats, and in places where its thickness continually diminishes occurs a transition from black to white—a perfectly homogeneous white tone. As soon as the coating dries, the relief contracts considerably; but the semi-transparency remains, and thus it is possible to reproduce the most beautiful half-tones of photography by printing. This relief-printing of Woodbury has already attained a high importance. It can work with any colour, and it facilitates the multiplication of photographic negatives by a single printing form, without the help of light. It is therefore of importance where a great number of pictures are required; for example, in reproduction of oil-paintings and drawings by the Relief Printing Company in London, Carbutt in Philadelphia; Goupil and Co. in Paris, and Bruckmann at Munich. Photographers do not use it much in portrait taking, because the production of a faultless gelatine relief and its impression on lead require a long process and an expensive apparatus, which would not pay in the limited sphere of portrait photography.

The frontispiece of the present work, representing the moon, from a photograph of Rutherford (mentioned p. 193), is an impression in relief, done by the Relief Printing Company in London.

It is a special advantage of the relief-printing process that it admits of printing on glass. Wonderful transparencies are thus obtained, very effective as window blinds. Goupil has prepared copies of oil-paintings in this style of relief, and they are frequently to be seen in the windows of our dealers. The transparent stereoscopic pictures on glass, produced by this process, are equally charming—in sharpness and softness they almost exceed the ordinary silver copies. Recently a number of beautiful magic lantern pictures, produced by the Woodbury press, have been offered for sale, and will be eventually used as an important means of instruction in schools. The author has a collection of American landscapes of this kind, which, when enlarged in the magic lantern, are more instructive than the fullest geographical treatise.

Pictures of this kind can be sold much cheaper than the ordinary transparent photographs for stereoscopes. (See the chapter on landscape photography, p. 163.)

SECTION V.—PIGMENT PRINTING, OR THE PRODUCTION OF CHARCOAL PICTURES.

Poitevin's Process—Production of Pictures in any kind of Pigment—Its Difficulty—Inverted Impressions—Transferring Process—Comparison of the Pigment Press with the Silver Press—Braun's Facsimile from MSS.—Transferrence of an Impression by Light through Pressure.

WE have seen above that gelatine mixed with chromate of potash is insoluble in the light. This fact was made by its discoverer, Talbot, the basis of heliography—that is, of photographic steel engraving. Poitevin, a Frenchman who has done much to promote photography, founded on the same method another process: he produced the pictures in different pigments (colours). He first used charcoal as a pigment, and he then obtained charcoal pictures.

The process is simple: Poitevin took gelatine coloured with printing ink, placed paper over it, and exposed this under a negative; then he washed the film of gelatine in hot water, by which means the soluble constituents perished, but the insoluble remained behind and retained the ink mixed with it; a charcoal picture was produced in this manner. Though this process appears very simple, yet it has its difficulties, as was remarked above in treating of photo-reliefs. The operation of the light often does not penetrate to the layer of gelatine film. The half-tones that have become insoluble do not therefore adhere firmly, and are detached in washing, as in Fig. 93 *e*, p. 234; therefore the pictures must be transferred before they are placed in hot water. This takes place as described at the end of the chapter on photo-reliefs. An albumenized sheet is pressed in the dark on the coloured film of the gelatine, and then the whole is plunged into hot water; on this the half-tones adhere to the paper pressed upon them, and the image appears uninjured on it, as in Fig. 93 *e*.

The position is, no doubt, reversed; that is, what was originally to the right in the lower image comes now to the left. It can easily be proved that this can be so. Let a word be written with thick ink, in large letters, on paper, and let a piece of blotting-paper be placed on the fresh writing, after which let the latter be removed; in this case the writing in ink has become printed off, but reversed. In the letter copying-press the same thing takes place, therefore the letters are printed on very thin paper, that they may be read on the reversed side, because viewed from that side they appear in their first position. Pigment impressions cannot be printed on such thin paper; therefore, if the reversed position is inconvenient, another transferring process must be adopted, by bringing the image from the first layer to a second. This end has been latterly attained in the following manner:—

The exposed gelatine surfaces are placed damp on a

smooth zinc table, and then suffered to dry. They then adhere very firmly to it. The copy thus glued to zinc is plunged into warm water, it is then developed, the paper becomes detached, and the image lies on the zinc table. A piece of white glue paper is now taken, stuck upon the zinc table, and allowed to dry. The image then adheres firmly to the glued paper, and being carefully loosened it parts from the zinc table and remains lying upon the paper. Then it appears in its proper position on the paper. In this more recent form the process is put in practice, especially at Woolwich Arsenal, in England.

The pigment impressions thus obtained resemble, externally, the Woodbury images, but they surpass them in their fineness and the facility of their production. But this process has not yet supplanted the silver press, for the expense of the material, owing to the twofold use of paper, equals that of silver photography, and the labour, being somewhat more complicated, is therefore dearer. The pigment impressions have a great advantage through the optional choice of colour; genuine Indian ink may be used for them, and then perfectly durable pictures are obtained that do not turn yellow or black.

In the same way English red, sepia blue, and so on, can be mixed with the gelatine, and thus pictures can be produced in those colours. This circumstance is important when it is wished to reproduce manuscripts, when these have been written in coloured characters. Quantities of such manuscripts and sketches of the old masters are in various museums. Braun of Dornach, in Alsace, the same photographer who made himself conspicuous for his

Swiss views, has undertaken to reproduce these manuscripts in their original colour by the pigment press, preparing first a silver negative in the usual way, and copying this on coloured gelatine films. In this way he has made accessible to all artists and amateurs, in faithful facsimiles, for a small sum, many drawings that only existed as a single copy.

Latterly, a very interesting observation has been made by Abney, in England, in relation to the pigment press. He remarked that if an exposed film of gelatine remained a long time in the dark the insolubility increases. Accordingly, a film of this kind which, freshly developed, would only give a faint image, after lying some hours gives a strongly-defined image. This fact allows the time of exposure for pigment pictures to be considerably reduced; that is, several pictures to be made at the same time.

Still more interesting is an observation of Marion, at Paris. He exposed a piece of common paper that had been made sensitive by plunging into a solution of chromium, then he pressed it in the dark with a damp sheet of pigment saturated with chromate of potash; the film of pigment was thereby insoluble in all places where it came in contact with the exposed parts of the chromic paper—it adhered closely to these parts of the chromic paper, and on developing with hot water he obtained a pigment picture on the chromic paper.

The future will show how far this process is practically useful, for there are always difficulties in practice, which require long experience to overcome them.

SECTION VI.—LIGHT-PRINTING.

The Susceptibility of exposed Chromic-Gelatine to the Influence of thick Printer's Ink—Services of Poitevin and Tessié de Mothay—Albert-type, or Light-press—Its Mode of Operation—Its Use and Comparison with Relief-Printing.

We have seen that a film of gelatine and chromium is insoluble in light, and loses its tendency to liquify. At the same time, the exposed places assume a peculiar property—they are susceptible to the influence of thick printer's ink. If an exposed sheet of gelatine and chromium is rubbed with a moist sponge, it only imbibes water in the places untouched by the light, but if it is then overlaid by thick printer's ink, it is singular that these places only adhere to the parts affected by light. This fact was discovered by Poitevin, the meritorious discoverer in photographic chemistry. If a piece of paper be placed on such a film of gelatine, blackened by ink, and it is pressed, the ink adheres to the paper and an impression of the image is obtained, under whose negative the film of gelatine had been exposed.

In this manner what is called a light-print is obtained. This peculiar mode of printing offered at first very imperfect results. The process was rendered unproductive from the fragile nature of the gelatine film, the difficulty of finding the right time for exposure, the proper consistency of the thick printer's ink, and other obstacles. After a hundred impressions, the film of gelatine was generally so injured that it was unserviceable. Tessié de Mothay, at Metz, acquired some skill in practising the process, but Albert, of Munich, was the first to develop it, so that it has become of practical importance.

The experimenters before Albert had conveyed the

gelatine film to metal, but it only adhered to this imperfectly. Albert poured the gelatine solution, decomposed with chromate of potash in the dark, on glass, and exposed its reverse side after drying for a moment. In this way the light produced a superficial effect, the part of the gelatine adhering immediately to the glass became insoluble, and was fixed wonderfully firmly to the glass. The film of gelatine was then covered on its upper surface with a negative, and exposed to the light. A faint greenish picture is thus produced. The exposed film is then washed in water until all salts of chromium is removed, and then it is suffered to dry.

In order to print, a sponge is moistened with water containing glycerine, and the film is carefully rubbed with it; the water penetrates in all places where the light has not operated. A leather cylinder is now taken and inked,—that is, some thick printer's ink is spread over a piece of marble by rolling the leather cylinder over it until the surface is coated,—then the gelatine film is subjected to a light pressure from the leather roller, which is passed over it, and this process is frequently repeated. All places which have been affected by the light receive ink from the roller, but not so the others, and finally a strongly-defined picture appears on the originally almost colourless surface. As soon as this has been sufficiently inked, a piece of paper is pressed upon it, and it is allowed to pass over a layer of gum, between rollers also coated with it. In this manner the ink of the picture passes over to the paper, and produces thus an impression with all half-tones. The inking and printing can be repeated at option, and thus thousands of copies may be prepared if the plate is very firm.

PLATE VI.

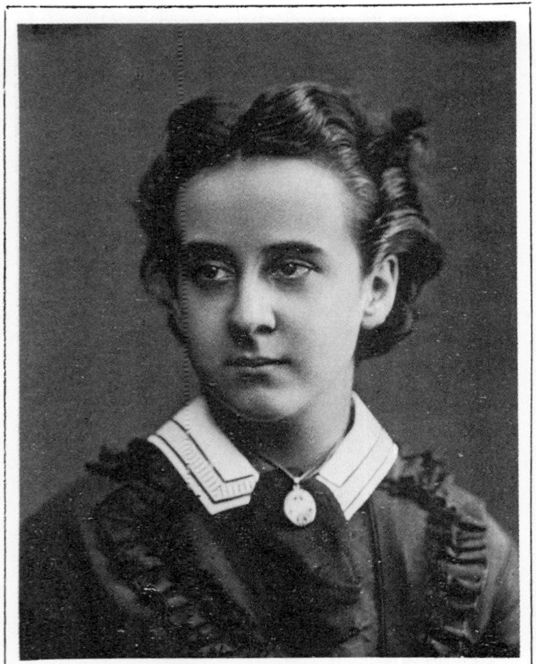

Copy of a Negative, untouched.

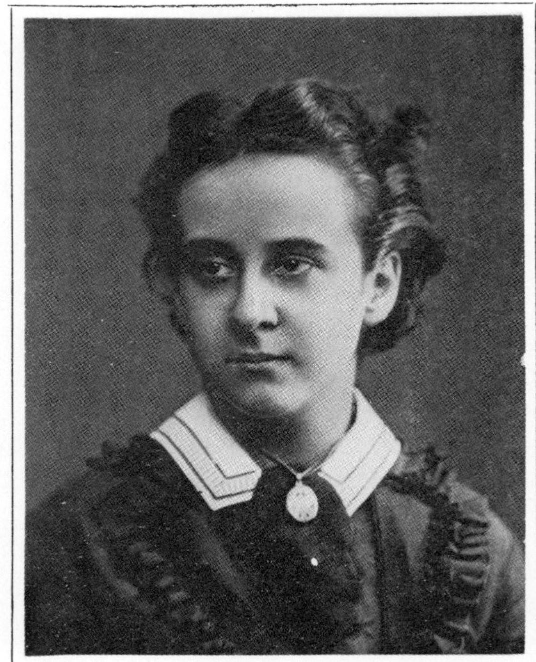

Copy of a Negative, retouched.

See pp. 50, 245.

These Albert-types, or light-impressions, as they have been latterly called, approach, but do not equal, the silver copies in beauty. The process is well adapted to reproduce pencil and chalk drawings, which they reproduce with the utmost fidelity. Herr Albert has reproduced and published "Schwind's Fairy Tale of the Seven Ravens," and several cartoons of Kaulbach, by light-impressions. In like manner, the views of the photographic detachment of the Prussian general-staff in the French war have been reproduced by Obernetter in light-impressions. The views of the Vienna exhibition, sold at the building, and by many supposed to be ordinary photographs, are light-prints of Obernetter of Munich, who with Albert has done the most in this matter. In the annexed double picture of Fraulein Artot we give our readers a specimen of a light-print of Obernetter.

The brilliancy of these pictures is occasioned by giving them a coat of varnish. If the results of the Woodbury printing are compared with those of light-printing, it appears that the relief-printing gives the shades and dark parts better, but that the white parts often appear discoloured. On the other hand, the relief-prints are much more like photographs than the light-prints, for the latter have a lithographic tone. It is only by coating with varnish that they are made more like photographs. But both methods are rather inferior to ordinary silver photography, which has never been surpassed in the uniformity of half-tones, in the beauty of lights, the uniformity of half-tones, and the depth of shadows, and which has one special advantage over light-printing and Woodbury printing, and that is the ease of production. To prepare relief and light impressions, the first thing needed is a printing plate, needing more complicated preparation than the photographic positive process, and also requiring a clever printer. But silver printing gives good results with simple means, and even in inexperienced hands. It will therefore always be preferred in portrait taking, where the object is often only to throw off a dozen pictures: but light-printing and relief-printing are of great importance when the object is to produce large numbers of pictures in a short time.

Section VII.—Aniline Printing.

Aniline Colours and their Origin—Effect of Chromic Acid on Anilinc—Its Use in Photography—Willis's Printing Process—Its Application.

Every one knows in modern times the brilliant aniline colours—Hofmann's violet, magenta red, aniline green, and others. We are indebted for these wonderful pigments, surpassing all before them in brilliancy, depth, and reflecting power, especially to the noted chemist, Hofmann. The colours are due to the effect of different substances giving out oxygen on aniline.

Aniline is a substance which resembles ammonia in its chemical relations, only it has a different odour and a different composition. The substance is obtained as a brown mass from coal tar on distillation.

If this brown fluid is treated with chloride or nitric acid, or manganese and sulphuric acid, or arsenious acid, different shades are produced. One specially interesting us is the colour which is created when aniline is heated with chromate of potash and sulphuric acid; the result is a peculiar violet substance—aniline violet. Chromate of potash plays a part in our photo-chemical processes, and on this is based the aniline-printing invented by Willis.

Willis plunges a piece of paper, in the dark chamber, in a solution of chromate of potash and sulphuric acid; he exposes the paper under a positive picture, *e.g.* a drawing or copper engraving. The light shines through the white paper, and in these places the chromic acid is reduced to chromic oxide, which does not affect aniline colours; on the other hand, the chromic acid remains unchanged under the black strokes, which keep back the light. After the exposure a very pale picture is seen of unchanged yellow chromic acid.

If this faint picture is exposed to fumes of aniline, at the places where the yellow strokes exist aniline brown is formed, and in this manner the original pale yellow becomes very defined. This fumigation with aniline is effected by placing the copies in a box and covering them with a cover, having a layer of blotting-paper on its lower surface moistened with a solution of aniline with benzine. This process reproduces a positive picture of a positive, and is therefore very valuable in producing faithful copies of drawings. I admit that these copies are reversed in position, as is the case with mirrors. This circumstance limits their use in many cases. We have already explained the reason of this reversing of copies (p. 240). Copies can, however, be obtained in their proper position if the original drawing is very thin. In that case the reversed side of the drawing is placed against chromic acid paper, and the light is suffered to shine through on it from the upper surface.

It is another disadvantage of this process that the chromic acid paper must be always freshly prepared, as it quickly spoils, and the duration of the copying must be very carefully estimated. If the time is too short, unchanged chromic acid remains everywhere on the paper, and then it all blackens in fumes of aniline. If, again, the time of copying takes too long, the light works gradually through the black strokes of the drawing, reduces the chromic acid, and the paper then remains entirely white in the aniline fumes, as no more chromic acid is present to form aniline colours. These circumstances limit the value of the method, and cause the *licht-paus* process (p. 25) to be preferred to it. In England the aniline printing is practised by the inventor, who prepares copies to order.

Section VIII.—Photo-Lithography.

Nature of Lithography—The Lithographic Colour-Press—Zincography—Poitevin's Discovery—Photo-Lithography—Its Application in multiplying Maps quickly—Its Importance in War—Difficulties—The Anastatic Method—Photo-Lithography with Asphalt.

By lithography is understood printing off from a drawn or painted stone. Near the little Bavarian town of Solenhöfen, there is a clayey, rather porous limestone, easily polished and worked. Such limestone is used for lithography. But the lithographic press differs essentially from copper printing and book press, because the drawing on stone is neither raised nor incised. The lithographic stone forms, in fact, with its image intended for printing, a smooth surface; and this process is peculiar, differing from all other modes of printing. If a drawing is made on a lithographic stone with chalk, or ink consisting of colour and a fatty substance, *e.g.* oil, and if the stone is moistened with water, it only penetrates the porous stone where there is no oily colour, for oil repels water. Oily printer's ink, similar to printer's ink, only considerably more delicate, is rubbed on the stone with a leather roller; it only adheres where there is the oily ink—that is, to the drawing.

After the stone has been inked as above, if a piece of paper is pressed upon it the ink passes over to it, and a lithographic impression is obtained. The stone can be evidently inked and used at discretion, and thus thousands of copies can be produced. This style of printing has many advantages over copper engraving. The working of a copper plate is a difficult matter, often requiring a labour of years to engrave, whereas the working on stone is much easier—almost as easy as drawing on paper. In like manner, printing from a stone plate has fewer difficulties than that from a copper plate. The stone easily admits of corrections in drawing, and after being used once, the same stone will serve again, and often for many years. These circumstances have brought lithography into general use: technical drawings, wine cards, pictures of saints, notes, visiting cards, lists of prices, calendars, illustrations of books, atlases, pictures of natural history, and a thousand other things are produced by lithography; and latterly a special field, called chromo-lithography, has obtained a large development. It is the most important of existing processes to produce painted pictures in a mechanical way. Chromo-lithography is rather more complicated than common lithography. If it is wished to make a chromo-lithograph of a painted picture, not only one stone, but a separate stone for almost every colour must be prepared. For example, if you have an object in which the blue, red, and yellow tones appear, you first draw on a stone that is to contain the blue places, and then colour it blue; a second and third stone are required for the yellow and red places. All three stones are printed off, in a position corresponding to the colours, on the same piece of paper. They then produce a painted impression which, if an oleograph is required, must be coated with a brilliant varnish. Though chromo-lithography offers great advantages for maps, ornaments, etc., and affords many excellent artistic specimens—*e.g.* the chromo-lithographs of Hildebrandt's water-colour views—we must express an adverse opinion against oleography, which, with a few honourable exceptions—Prang at Boston, Korn at Berlin, and Seitz at Hamburg—produces pictures of very small artistic value, and has done much to injure the public taste.

A perception of colour and a feeling for art are necessary for chromo-lithography, and the printers do not possess them.

Closely related to lithography is zincography, which we shall glance at here before passing to photo-lithography.

It is remarkable that zinc has properties similar to those of the lithographic stone. It easily receives drawings with soft chalk, and after being moistened with gum water, it can be as easily rolled as a stone with moist colours, the colour adhering then to the places drawn. The printing, therefore, presents results similar to those of lithography; but the preparation for zinc printing has more difficulties than lithography, so that the use of zinc for this purpose is limited.

We have only given a brief survey of the principles of

lithography and zincography, as far as necessary to understand what follows. Both processes resemble light-printing in many respects; the light-press surface has also the peculiarity of receiving thick ink in some places and repelling it in others,—but light-printing is of recent date, while lithography has existed seventy years. When photography was invented, it deprived lithography of an important branch, that of portraits. Even in 1850 numerous lithographic portraits were made of individuals. But since the introduction of cartes de visite at that date, portrait lithography has greatly fallen off, and is only used for cheap likenesses of distinguished persons. The lithographs from oil-paintings have also suffered through photography, which thus entered into competition with lithography. It was Poitevin who allied the two by inventing photo-lithography. It was Poitevin's aim to economize the labour of the lithographic draughtsman, and to replace it by the chemical operation of light. He coated lithographic stones with chromate of potassium and gelatine, and exposed them under a photographic negative. The chromo-lithograph thus obtained was then washed and rolled. All parts affected by light took the colour, and gave an impression in the press. The first attempts of the kind were very imperfect; the pictures were especially wanting in half-tones, which were lost in washing, as they are in the pigment press (see p. 239); therefore Asser and Osborne attempted another manner, what is called re-impression. They copied their negatives on chromic paper, partly coated with gum gelatine or albumen, and then they rolled it. Chromic paper has the peculiarity of receiving thick printer's ink on the exposed places after exposure. After inking, the chromic paper was carefully washed and then pressed on a lithographic stone. This sucked up the colour, and thus the picture was perfectly transferred to the stone. The stone thus prepared gave excellent impressions in the usual style of lithography. Though half-tones were thus produced, the impression fell far short of photography in quality. The lithographic half-tone differs essentially from the photographic, which forms a homogeneous surface, while the lithographic appears as a mass of black points more or less near together. The granulous nature of the stone does not allow such delicacy of reproduction as photography; therefore, photo-lithography is employed only where cheap production of many copies is of greater moment than delicacy.

Quite recently, maps of rivers and mountains have been sold by Kellner & Co., at Weimar, which have been produced by photographing reliefs in gypsum, and copying the negatives obtained by photo-lithography. They thus succeeded in producing maps of mountains without the help of a draughtsman, and could sell them at an extremely low price.

Such photo-lithographs do not answer as works of art. In this sphere light-printing, with its excellent half-tones, is a formidable competitor, though photo-lithography is very useful by giving a great number of impressions from the same plate, while the number of light-prints given off by a gelatine plate is always limited and rather uncertain.

In one branch photo-lithography surpasses all other reproducing arts; that is in giving copies of maps, which have first been drawn in outline. The production of geographical maps is a field requiring much time and observation. The individual outlines of mountains, rivers, and countries must be executed with the greatest exactitude, corresponding with measurements. Frequently, different draughtsmen and engravers are employed, and, though working conscientiously, inaccuracies are unavoidable and make correction necessary. All this takes time and trouble. If the object is now to take an enlarged or diminished copy of a map, the same difficulties occur, and the diminishing is especially troublesome. The pantograph is a useful aid here, but does not exclude inattention in the draughtsman. In this respect photography is invaluable as an aid to map-making. With very great ease, photography gives enlarged or diminished copies of an original; in a few hours this is copied on stone, and within a day photo-lithography can throw off thousands of enlarged, diminished, or original sized copies.

If it were wished to make a lithographic stone drawing by handiwork, several days would be wanted, and it would be far less exact. No photographic printing process is as rapid as photo-lithography; therefore map-making has made great use of it, especially when very numerous copies were required. In the French war, the advancing troops needed, before all things, maps of the territory to be occupied. But there was not a sufficient supply of maps of France to provide whole Army Corps with them. It was not to be thought of before the war, as no one can tell where a campaign will take place. Photo-lithography was here an auxiliary, by giving thousands of copies from one original map; and thereby it contributed to the successful advance of the German army, which, with these maps in its hand, showed itself better acquainted with the enemy's territory than the enemy's troops themselves. The photo-lithographic establishment of the brothers Burchard, at Berlin, which has displayed activity on a large scale, produced in the war time of 1870-71, 500,000 maps. Plate V. is a specimen of this process.

Besides these contributions, we must mention the photo-lithographic labours of Herr Korn at Berlin, which belong more to the region of art. Particularly admirable in this branch are the photo-lithographic pages of the pen-and-ink drawings of Berg in the Japan expedition. These are so faithfully produced that original and copy cannot be distinguished,—I admit that the character of the originals was very favourable to photo-lithographic reproduction. Faint drawings place difficulties in the way of photographic reproduction, especially when they have a bluish tint; therefore, pencil drawings are very difficult to photograph. But no perfect photo-lithograph can be produced from an imperfect photographic negative. Thus far, therefore, the character of the original has a very material influence. Berg's pen drawings are executed in strong black Indian ink, and therefore easy to reproduce. In the Austrian Institution of Military Geography, the map drawings which are to be copied by photo-lithography are so executed from the first that they take a favourable photographic effect, or, to use the technical term, come out well. The photographic reproduction of a drawing is especially favoured by brown tints, such as umber and cinnabar, when mixed with Indian ink. On the other hand, much depends on the paper being without

REDUCTION OF
A SECTION OF THE ORDNANCE SURVEY MAP
OF WALES

PLATE V.

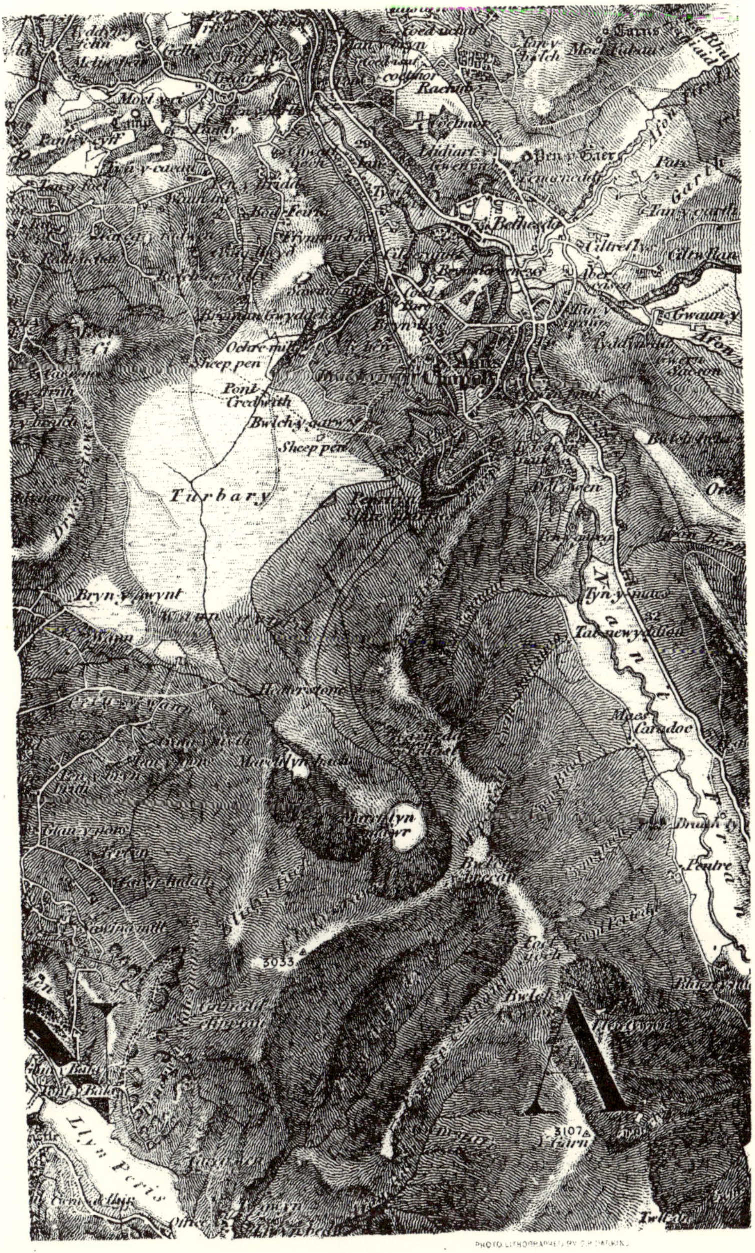

blemish; yellow spots, scarcely visible to the eye, have the same effect in photography as black. We knew a case in Kron's photographic printing-office where an unblemished drawing of a map came out as a photograph full of spots. The defect was attributed to the chemicals, until it was found that minute rust spots in the paper, which had got into it during manufacture, were the cause of the defect. In such cases the evil can only be remedied by suitable negative *retouche*.

The nature of photo-zincography will now be clear to the reader: as the zinc plate is so like the stone, the treatment is the same. The negative is either copied direct on the zinc plate, coated with gelatine and chromium, or a copy from the negative is prepared on chromo-glucose paper; the paper is inked and transferred to the zinc plate, being pressed together with it. After this the zinc plate can give impressions.

It must be remarked in this connection, that even without photography, direct mechanical copies can be made of maps, writings, etc., if the original be executed in oily or analogous colours. This takes place by means of the anastatic process. This process is based on moistening the original on the reverse side of the drawing with acidulated gum-water, and then damping it from above with a fresh colour; this only adheres to the oily strokes of the drawing or printing. The original, thus freshly inked, is then placed on a fresh stone, or a freshly cleaned zinc plate, and put under a pressure. Then the drawing passes over to the stone or the zinc, and can be easily multiplied by rolling and printing. It is difficult to preserve the original, which suffers greatly under pressure. Still more difficult is it to reproduce a pure stroke, for these are often extended widely by the pressure, and if the strokes are too thin they run together, as in mountain lines in maps; therefore the process has been more applied to copy antiquarian books, which have been reproduced page by page in this way.

It is self-evident that only reproductions of the original size can be made by the anastatic process.

We have to mention another process of photo-lithography, based on the application of asphalt. We have already described this in our first chapter as a sensitive substance, and also a process called heliography, which produces, by means of photography, copper plates and steel plates for printing. Asphalt serves also for photo-lithography. A lithographic stone is sprinkled with a solution of asphalt in ether, allowed to dry in the dark, and exposed under a negative. The asphalt becomes insoluble on the exposed places, and is retained upon treating the stone with ether or benzine. If the stone is then damped, the moisture only penetrates where no asphalt covers the stone. On rolling it after this with oily ink, this is rejected from the damp places, and only adheres to the asphalt—that is, to the picture; thus a stone giving impressions is obtained. This method gives good results in the hands of several practitioners, and is preferred by many to the chromium process, though asphalt is much less sensitive than chromium.

SECTION IX.—PYRO-PHOTOGRAPHY WITH SALTS OF CHROMIUM.

Poitevin's Process—Effect of Chromate of Potash on Sticky Substances —Pictures Developed by Dust—Pictures on Porcelain—Oidtmann's Pyro-photography—Application to the Decoration of Glass—Photography and Painting on Glass.

Photography has become allied to almost all the multiplying and descriptive arts, though it was at first looked upon as their rival. It is not surprising, therefore, that it has become a help in porcelain painting and decoration. We have already seen (p. 212) the peculiar process of changing silver pictures into gold and platinum pictures, transferring them to porcelain, and burning them in. That method might be called a moist process; the same end can be obtained by a dry method, and by the help of salts of chromium. This original method has also been invented by Poitevin, and subsequently was materially improved by Joubert in London, and Obernetter at Munich. It consists in this: that a mixture of gum, honey, and chromate of potash is poured on glass; the film is carefully dried in the dark, and then exposed under a positive. The film of gum is freshly prepared, sticky, and holds fast the scattered coloured powder, but when the film is exposed, it loses its stickiness. If this exposure takes place under a drawing with black strokes, the film under them will retain its stickiness, but lose it beneath the white, transparent parts of the paper.

Therefore, if the film, after exposure, be powdered in the dark with any colour in powder, this adheres where the strokes of the drawing have protected the film, but not at other places, and thus a picture in powdered colour is obtained. If this coloured powder and its underlayer is fire-proof—as glass and porcelain—the picture obtained can be burnt into it, and pictures of very different shades can be produced, according to the choice of the powdered colour. Pictures of this kind can be transferred from one under-layer to another. If a collodion film is poured upon the powdered picture, if this is suffered to dry and then the whole thrown into water, when there the collodion film with the picture can be easily taken off, stuck on, and burnt into other surfaces—glasses, cups, etc. Thus Joubert in London has actually burnt in large pictures on glass. Obernetter at Munich, and Leth at Vienna, Leisner at Waldenburg, and Stender at Lamspringe, Greiner at Apolda, and Lafon de Camarsac at Paris have produced encaustic pictures on porcelain in the same manner.

This process is not much employed for portraits. On the other hand, Oidtmann at Linnich,* has employed it advantageously in glass manufacture. He has copied patterns of carpets from lithographs directly on glass, and burnt them in, thereby producing cheap window ornaments, which can be painted and embellished by the hand. At the Vienna exhibition, there was over the door of the Emperor of Germany's pavilion a rosette ten feet in diameter, produced by Dr. Oidtmann on the above system. The same person has also employed the process to produce mosaic glass pictures, similar to the mediæval glass paintings on glass. These mosaic glass pictures are produced by cutting out coloured pieces of glass corresponding to the figures and their

* See "Photographische Mittheilungen," Jahrg. 1869. Berlin: Oppenheim.

colours. For example, for a human figure, the outline of the face was drawn on a flesh-coloured glass slab and cut out; the same thing took place for the drapery, on glass slabs corresponding to the colours. The lights and shades and details—for example, nose, mouth, and eyes—were then drawn with black moist colour on the proper piece of glass assigned to it, and burnt in, after which all the separate pieces of the glass were soldered (or cemented) together. Dr. Oidtmann does, by means of photography, what the draughtsman does in this mosaic glass-painting. He copies the outlines of the face from the large-sized original photograph, or the original woodcut, on the proper piece of glass, and powders it with moist black paint, and he thus obtains a picture that can be burnt in, and which can be treated in the manner described. At the Vienna exhibition there was a copy of "The Crucifixion" by Dürer, produced in this manner, and composed of 150 glass pieces. Dr. Oidtmann prepares the large-sized original pictures by magnifying little woodcuts according to the photographic manner (see p. 95). Dr. Oidtmann has also attempted to produce pyro-photographs, by proceeding on the principle of chromo-lithography (see p. 250, Photo-lithography). He copied the similarly coloured parts of a painted drawing—covering over the others—on a film of gum and chromium, powdered this with a suitable colour, and then copied the other colours of the original successively in the same manner. He thus obtained a powdered picture, which was then burnt in.

SECTION X.—PHOTOGRAPHY AND THE SAND-BLOWING PROCESS.

The Nature of the Sand-blowing Process—Its Connection with the Pigment Press—Its Employment in Heliography instead of Corrosive Acids.

Tilghmann, at Philadelphia, made the observation, during his residence at the watering-place Longbranche, that the windows exposed to sea wind became quickly misty. He found that this was occasioned by fine sand, which the wind drove against the window; this gave him the idea of making ground glass by sand blown on to it, and this succeeded perfectly. He covered a glass surface with an iron mould, in which figures and letters were cut; he kept this in a current of air bringing sand with it. In a short time this made ground glass at the places that were exposed, and Tilghmann obtained thus a drawing of the incised figures. A blast of only four inches hydraulic pressure and a period of ten minutes are required for this work. If the air pressure is stronger, or steam is used instead, conveying sand, and having a pressure of 60 to 120 lbs. to the square inch, the effect is immense. Sand blown with such power through a narrow pipe bores deep holes into the hardest stones, and even into glass. The process has been used to bore stone and metal plates. If a mould of cast-iron is placed on it in which the figures have been cut, they can be deeply engraved in a short time in the stone. The iron plate is, no doubt, injured, but much more slowly than the stone slab. A cast-iron plate $\frac{3}{16}$ of an inch thick is only reduced $\frac{1}{16}$ of an inch, whilst a section 300 times deeper is made in marble. India-rubber endures the sand stream almost as well as iron. You might cut into marble with an india rubber-mould 200 times as deep as the depth of the mould, without much injuring it.

With the pressure of 100 lbs., such a sand stream can penetrate 1½ inches deep into granite, 4 inches into marble, and 10 inches into soft sandstone.

The circumstance that soft bodies act as shields in this has led to elegant applications of this method in the industrial arts. For example, if glass be covered with lace pattern and a sand stream take effect upon it, the glass becomes ground in the meshes, and a copy of the lace is obtained on glass. In the same manner, you can paint with gum colour upon glass, and this drawing can be produced clear on an unpolished ground by the sand blast. This circumstance led immediately to the application of photography. If a pigment impression (p. 239)—that is, a chromo-glucose picture—is produced on glass by transferring a prepared impression directly on glass (see above), the glass surfaces at all the places of the picture are protected by a layer of gelatine. If now a sand stream is allowed to operate upon it, it polishes the glass only at the uncovered places; thus an opaque and transparent glass picture is the result. If the gelatine picture is a negative, the shadows are dim; and such an unpolished slab is also fit for giving impressions by means of printer's ink. Corroding with acids often eats into the fine strokes and makes them broader. In place of them the heliographic metal plates of Talbot (p. 225) can be blown upon with sand—which, owing to its perpendicular position, only works downwards—and thus cavities of great depth can be produced, so that plates thus blown upon can be used for high relief-printing, that is, book printing. Tilghmann recommends that a positive of gelatine and chromium should be produced upon a cake of resin; that this should be blown upon and deeply hollowed out; then a form is obtained which can be first cast in gypsum, and then in metal type, which can be used for printing.

These are interesting experiments, which ought in time to lead to important practical results.

Section XI.—The Photometer for Chromo-Photography.

In many of the above described chromic processes, *e.g.* the production of relief-prints, pigment-prints, light-prints, etc., it is very important to determine the exact time for exposure. This is not easy, because the picture

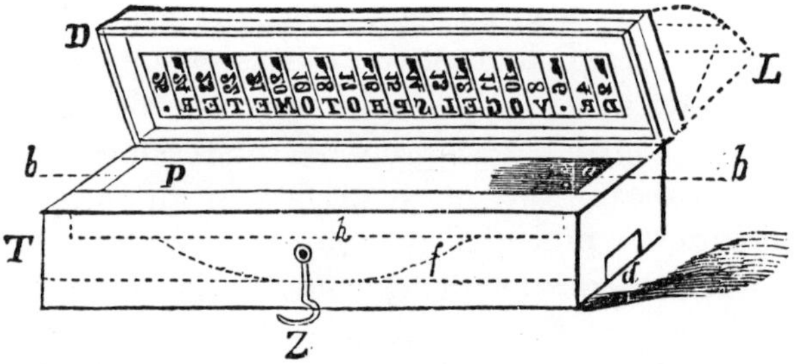

Fig. 94.

appears only faint or not at all, as in pigment-printing; therefore the state of the picture gives no safe criterion respecting the completion of the picture. This circumstance has necessitated the application of a photometer, easily determining the duration of the exposure. Such photometers have been provided by Byng and Swann in England, and by the author. The author's photometer consists of a semi-transparent paper scale *L*, whose transparency diminishes from 2 to 25. (See Fig. 94.)

This scale is formed of layers of paper, whose entire

quantity is expressed by the figure printed on them; under this scale is exposed a strip of chromic paper; that is, paper which has been plunged into chromate of potassium. The strip is enclosed in a little box in such wise that, when the cover *D* is shut down with the scale, the chromic paper and the scale are in close contact; the light now shines through the scale and browns the paper strips lying under it. This colouring affects, first, the thin transparent part of the scale, and passes thence to the opaque end, the rapidity depending on the strength of the light. To know how far the effect of the light has extended, figures are printed on the scale which do not permit the light to pass; therefore these remain clear on a brown ground, and the place to which the effect of light has advanced is perceived by the figure that appears there.

To use this instrument, some experimental copies must be made first. Supposing it were desired to prepare a pigment impression from a negative, the film of pigment is exposed under the negative at the same time with the photometer. After some time, lamplight is used to see how far the photometer paper is browned. The significant number is noted—photometer degree—and the negative is only half covered, the other half continuing to be copied until a higher degree of the photometer. Then the pigment picture is developed, and the degree of the photometer is determined where the favourable result has been obtained. Rarely more than one attempt has to be made; when this has determined the degree up to which the negative must be copied, the time of exposure can always be regulated with the help of the photometer. Practised hands only determine the degree with some negatives, and easily ascertain up to what degree a fresh negative must be copied.

SECTION XII.—THE CHEMICAL EFFECT OF LIGHT AND THE PEA-SAUSAGE.

In the campaign of 1870, the well-known pea-sausage was one of the most important articles of food for the army, and was prepared daily in many thousands of skins. The fabrication of the interior portion caused little difficulty, but the obtaining so many skins created much difficulty. As the supply fell short, a substitute was sought in parchment. This paper, which is produced by dipping for a second blotting-paper in sulphuric acid, then washing and drying it, is distinguished by its skin-like properties of resistance. It is impenetrable to water, and difficult to tear. It is therefore used for the production of cheques on the Treasury. It was attempted to fabricate sausage skins with this paper, by doubling a sheet cylindrically and pasting it together. No glue or gum can resist the effect of the boiling water in which the sausage has to be cooked, and so the artistic sausage skin fell asunder. Dr. Jacobson solved the problem by producing an adhesive substance, with the help of the chemical effects of light, which could resist boiling water. He mixed the gelatine intended for the pea-sausage skin with chromate of potash, and exposed the adhesive parts to the light. This occasioned the insolubility of the gummy substance, and now the artificial skin endured boiling water thoroughly well. The number of sausage skins prepared in this way, by the chemical operation of light, amounted to many hundred thousands.

CHAPTER XVI.

IRON, URANIUM, AND COPPER PHOTOGRAPHY.

Historical—Combinations of Iron—Effect of Ether on a Solution of Chloride of Iron—Chloride of Iron and Paper—Iron Pictures in Blue —Iron-gold Pictures—*Paus* Process with Salts of Iron—Iodide Pictures—Combinations of Uranium—Uranium Pictures—Their Development—Copper Pictures of Obernetter.

WE remarked further back that the number of sensitive substances is much greater than appears, and a close analysis was to determine that all bodies were more or less sensitive to light. Even in the first period of photography, 1840, Herschel observed the sensitiveness of salts of iron, Burnett that of salts of uranium, and Kratochvila prepared successful daguerreotypes on copper plates, in a manner analogous to those on silver plates. This process has been energetically cultivated, but hitherto without any important result.

It has long been known that chloride of iron, a yellow substance consisting of iron and chlorine, when dissolved in ether bleaches in the light and becomes the hypo-chloride of iron, having less chlorine. The same thing takes place in connection with paper. If clean paper is saturated with a solution of chloride of iron in six parts of water, dried in the dark, and exposed under a negative picture, the paper, which is yellow at first, becomes white under the transparent places, because the yellow chloride of iron passes into white hypo-chloride of iron. This pale hypo-chloride picture can be easily coloured intensely dark. If the pale picture is plunged in a solution of red prussiate of potash, this, combined with the hypo-chloride of iron reduced by light, easily produces Berlin or Prussian blue, while it leaves the chloride of iron unchanged; in this manner a blue picture is obtained. If a pale iron picture be plunged in a solution of gold, it becomes of a light blue colour, because the hypo-chloride of iron produces a precipitate of metal gold. In this manner a dark precipitate, which is produced by hypo-chloride in many substances, will give a dark colour in all such cases.

Another process is the transforming of iron into iodide pictures. A piece of paper, saturated with chloride of iron, is copied under a positive (for example, a drawing). The copy comes out as a yellow drawing of an unchanged chloride of iron on a white ground. If the paper be now plunged in a solution of iodide of calcium and starch, the iodide becomes liberated, and forms with the starch a dark blue iodide starch, which gives a strong dark shade to the lines that were originally pale (Herschel).

There are several other processes to make iron pictures of a darker colour. The pictures in Berlin blue do not hold, because that colour turns pale in the sun (accordingly, blue parallels rapidly lose their colour in the light). The same remark applies to pictures of iodide of starch; the gold pictures are too pale and their restoration too costly.

The salts of uranium present the same phenomena as these salts. Uranium itself is a rare metal whose combinations play a great part in colouring materials; thus there is a yellow oxide of uranium, that can be burnt into porcelain, giving a dark green colour, and which, being mixed with glass, imparts to it a beautiful grass-green (Anna glass). Moreover, a chloride of uranium and a hypo-chloride of uranium are known, corresponding to the chloride of iron and the hypo-chloride of iron, and bearing much resemblance to the combinations of iron just mentioned. The most noted salt of uranium is nitrate of uranium, which is reduced to sub-nitrate of uranium by the influence of light in the presence of organic bodies—for example, paper receptacles. If a piece of paper is dipped in a solution of one part of this salt and five parts of water, if it be then dried in the light under a negative, a very faint, scarcely perceptible picture is obtained, consisting of sesqui-oxide of uranium. If this is plunged in a solution of silver, or a solution of gold, it becomes suddenly visible, because the sesqui-oxide of uranium precipitates directly the gold or silver metal as a coloured powder (in the case of the silver, brown; and in that of gold, violet).

The uranium is too rare and too dear to be employed generally in photography.

As can be perceived, the salts of iron and the salts of uranium are analogous to the salts of chromium, by only being sensitive to light in the presence of organic bodies. In a pure state, salts of uranium and salts of iron do not change in the light.

The sensitiveness to light of salts of copper has hitherto only been studied very imperfectly. Copper forms with chlorine a green salt, soluble in water,—chloride of copper,—which is reduced to hypo-chloride of copper in the light. Obernetter took advantage of this fact, mixing chloride of copper and chloride of iron together, and saturating the paper with them. This was exposed to light under a negative, then plunged in sulpho-cyanide of potassium, and ultimately treated with red prussiate of potash. The result produced by this somewhat complicated process was a brown picture.*

* See Vögel, "Lehrbuch der Photographie," p. 32. Berlin: Oppenheim.

CHAPTER XVII.

THE CHANGE OF GLASS UNDER THE INFLUENCE OF LIGHT.

Faraday's Observation on Manganese-glass—Change of Mirror-glass in the Light—Almost all kinds of Glass are Sensitive to Light—Gaffield's Experiments—Disadvantages of the Change of Glass in the Light—Explanation of the Change of Manganese-glass—Operation of Light on Topaz.

THE celebrated natural-philosopher Faraday made the observation that glasses painted with manganese, and conspicuous for a peculiar flesh tint, became rapidly brown in the light. This fact remained for a long time without further results. But some years later, other observations of the same kind were made.

A very handsome plate of glass was exhibited in a mirror shop at Berlin. It bore the inscription "Manufacture of Mirrors" in brass letters. After being exhibited for years, the business was broken up, and the mirror, on account of its beauty, was taken away by its proprietor, the brass letters were effaced, and the plate was cleaned. To the surprise of the proprietor, the letters remained plainly visible on the glass, notwithstanding all attempts to remove them. The surface was even abraded, but this did not produce any effect on the letters. It was found that the glass was penetrated with yellow marks, and that it remained white only at the places where the opaque letters had kept off the light. The plate of glass was afterwards cut into two halves. One half, with the word "mirror," remains in the hands of the Philosophical Collection of the University of Berlin. Gaffield has lately made very interesting observations on the change of glass in the light, and has thus determined that almost all kinds of glass are sensitive to the light, and that often an exposure of only a few days suffices to effect this change.

Gaffield went systematically to work in his experiments. He cut the glass in question into two parts, placed one in the dark and the other in the light, and compared the two after a few days. In almost all cases he remarked a darkening of the colours. Only two kinds of greenish German and Belgian window-glass remained unchanged. The glasses of a darker colour lost their colour when exposed to red heat.

This change of glass in the light has a very unfavourable effect in photographic studios. Through the yellowish colouring which their glass assumes, in time a part of the chemically operative light is absorbed in the glass. The deterioration of light thus resulting makes itself remarked in a very conspicuous manner, because the time which is necessary in order to take a portrait must be continually lengthened.

The glasses that contain manganese change most strikingly. Hyper-oxide of manganese, also named brownstone, is often added to glass to discolour it. The dark green sesqui-oxide of iron in the glass is transformed, by the operation of the oxygen of the brownstone, into the paler protoxide of iron, and the discolouration is effected in this manner. The opposite effect takes place in the light. The oxide of iron is reduced again to sesqui-oxide of iron; the oxygen passes to the manganese, and, forming brown oxide of manganese, gives rise in this manner to the dark colouring.

In many minerals, light has an opposite effect to that which it has on glass. It does not colour them, but discolours. This happens especially with the Siberian topaz, which soon loses its golden-yellow colour in the light. A splendid crystal of topaz, six inches high, belonging to the Mineralogical Museum at Berlin, has in this manner lost materially in the beauty of its appearance.

CHAPTER XVIII.

PHOTOGRAPHY IN NATURAL COLOURS.

Observation of Seebeck and Herschel—Bequerel's Painted Pictures and Silver Plates—Nièpce's Labours—Effect of Black Colours—Coloured Pictures on Paper of Poitevin and Zencker—Want of a Fixing Medium for Coloured Photographs.

Photography has already achieved grand results; but it has still one problem to solve—the production of photographs in natural colours. There are plenty of coloured photographs to be seen, but in such cases the colour has been added after by the paint brush; it is a kind of *retouche*, which in most cases does not improve the picture. But we are now speaking of photographs in natural colours, reproducing the original colours of objects solely and alone by the operation of light. Numerous attempts are at hand, pointing more or less to this great end. The production of coloured pictures, by the chemical effect of light, has been successfully achieved; but these are spoiled soon by the influence of the same agent to which they are indebted for their production. No means exist at present of fixing coloured photographs.

The first attempts to make coloured pictures date a

long way back. Professor Seebeck of Jena, as early as 1810, found that chloride of silver took in the colour spectrum almost the same colour corresponding to them. This observation, published in Goethe's "Farbenlehre," II., page 716, passed unnoticed. Only since 1841, after the discovery of the daguerreotype, the noted Sir John Herschel made experiments in the same direction. He took paper saturated with chloride of silver and nitrate of silver, let a powerful solar spectrum fall upon it, and obtained immediately, like Seebeck, a coloured image of the spectrum, agreeing, however, only imperfectly with the original. Bequerel was more successful. He ascertained that the solution of nitrate of silver in Herschel's experiments had a disturbing effect, and he worked with chloride of silver alone. He employed silver plates, which he plunged in chlorine water. The plates become thus whitish, by the formation of chloride of silver, and, when exposed to the spectrum, present a picture whose colours agree very nearly with the natural ones. Bequerel observed that the duration of the operation of the chlorine water was very important, and he preferred at a later date to chlorize the plates by the operation of the galvanic current. To this end he suspended them to the copper pole of a galvanic battery (see p. 227) and plunged them in salt acids. The galvanic current decomposed these acids into chlorine and hydrogen. The chlorine passes to the silver plate, and forms chloride of silver. This method enables the operator to produce a film of chloride of silver of any thickness, according to the duration of the operation of the electric current. The brownish hypo-chloride of silver is thus produced, and this is chiefly sensitive to colour. Yet this sensitiveness is not great, it only suffices to fix a powerful spectrum, but it requires a very long exposure to obtain images from the camera obscura, and all such images, unfortunately, darken through the continuous operation of light. Bequerel found that the sensitiveness was increased by heating the plates. This observation was turned to account by his successor, Nièpce de St. Victor (the nephew of Nicophore Nièpce, see p. 9). This savant made numerous experiments from 1851-67 to produce coloured photographs, and he imparted his observations to the Paris Academy.

He worked, like Bequerel, with silver plates, which he chlorized by plunging in a solution of chloride of copper and chloride of iron, then heated to a high degree, and thus obtained plates which appeared ten times more sensitive than Bequerel's, and allowed him to copy in the camera obscura copper lithographs, flowers, church windows, etc. He relates that he not only obtained colours, but that gold and silver remained in their metallic splendour in pictures, and the picture of a peacock's feather retained the lustre of nature.

Nièpce de St. Victor introduced a further improvement, by covering the plate of chloride of silver with a peculiar varnish, consisting of dextrine and a solution of chloride of lead. This coating made the plates still more sensitive and durable. At the Paris exhibition of 1867, Nièpce exposed the different coloured photographs that lasted above a week in a subdued daylight (they were exposed in half-closed boxes).

Among these pictures were a couple of uncoloured, but perfectly black pictures on a white ground, which had been copied from copper-plate engravings. These excited great interest, and justly so, because in these pictures the darkest influence had had most effect, and the cleanest and whitest the least. This was, therefore, a directly contrary effect to what happens on photographic paper, where the dark produces a clear effect, and the clear a dark effect (see p. 28). This production of black by black can only be explained by assuming that the black is actually not black, but that it irradiates ultra-violet light invisible to the eye (see p. 60).

Since Niepce, who died in 1870, the only persons who have directed attention to coloured pictures are Poitevin at Paris, Dr. Zencker * at Berlin, and Simpson in London. But the two former investigators have reverted to the older process, as employed by Seebeck and Herschel, *i.e.* they prepared pictures on paper again, only the preparation of this paper was peculiar. Paper saturated with salts was made sensitive in a solution of silver, like the photographic positive paper (see p. 50), then washed to remove the solution of silver, and afterwards exposed to the light in a solution of hypo-chloride of tin. By this means violet hypo-chloride of silver is formed from the white chloride of silver. The hypo-chloride of tin only operates as a reducing medium. This paper is in itself little sensitive to colouring; but if it be treated with a solution of chromic acid of nitre and copper vitriol, its sensitiveness increases considerably, so that it is easy to copy with it pictures of transparent colours. Nevertheless, the colours are never so vivid as in the original, the red tones showing themselves the

* Those who take a special interest in the matter are referred to Dr. Zencker's "Lehrbuch der Photochromie." Berlin, 1868.

strongest. After copying, the pictures are washed with water, to make them less sensitive to light. In this condition they showed tolerably well in a subdued light, but no means have been found hitherto to make them perfectly durable. The fixing natrium of the photographer (see p. 27) cannot be employed, as it destroys the colours directly. We must hope that future investigators will succeed in supplying this deficiency. The first attempts in uncoloured photography also failed for want of a fixing medium (see p. 6), which was only discovered seventeen years later by Herschel.

CHAPTER XIX.

PHOTOGRAPHY AS A SUBJECT TO BE TAUGHT IN ART AND INDUSTRIAL SCHOOLS.

Importance of School Photography—Its Use for Technical Institutions—Photography as an Object to be taught in Art Schools and Universities.

THE previous chapters prove how manifold are the applications of photography. It has entered into art, science, industry, and life as a new kind of written language. Photography is to appearance what printing is to thought. Typography multiplies what is written, photography what is drawn; nay, more, it also draws in a chemical way. No doubt a certain technical practice of the art is required that can only be gained by experience; but it is easy to learn, and the time cannot be distant when it will be taught as an extension of drawing—itself a matter of tuition—in all industrial schools. Years are devoted to the study of the art of drawing, of piano playing, and other things; a course of instruction lasting half a year would suffice to learn photography.

The author has for nine years presided over a professor's chair of photography in the Royal Industrial

Academy at Berlin, the only technical institution in Germany where photography has as yet obtained a footing in the curriculum. It is by no means the object of this institution to train professional photographers; it only admits photography so far as it has importance for industry and science.

At this institution practical exercises are carried out in the positive and negative processes of photography, especially in its application to reproduce drawings, for taking machines and buildings; and, moreover, instructions are given on the *licht-paus* process. Other technical institutions are still hesitating about admitting photography. The importance of the matter is still depreciated, and what is new is still viewed by many as inconvenient. We cannot avoid introducing, in this connection, a passage from a work that has recently appeared: "Photography as a Matter for Teaching in Schools of Industry," by Professor Krippendorf, of Arau. The author remarks:—

"Schools of industry are instituted with the special view to prepare pupils for the subsequent professions of a technical and industrial life, and therefore they naturally admit in their curriculum the arts and sciences devoted to this end, especially drawing. Industrial training must see not only that these branches form an organic whole, taking the place of the old classical languages as a basis of general culture; it must also draw into the province of science the new inventions introduced in industry, and then suffer them to work back on practical pursuits in a more profitable manner.

"Photography is one of the branches that has advanced most rapidly in the last ten to twenty years. It is a genuine product of natural science, not as the mere plaything of accidents, but having the great merit of being first conceived as an ideal, and then practically carried out. It is therefore an art full of value in itself, based on science, and one whose productions delight and are gladly viewed by all, extending the knowledge of pupils and giving an idealizing tendency to young minds.

"There is scarcely any other branch in industrial schools in which it is a downright necessity to keep in view an independent observation of the result. Physics and chemistry are taught as successful results, and give no clue to detect the source of errors. The pupil only observes what the teacher puts before him, and both are satisfied if the law is found in the experiments. A learning how to observe does not properly take place, yet this is specially fitted to sharpen the judgment. But if photography is admitted in the school, we gain a branch which fixes the attention of the pupil in a way that no other can do. The study of photography is specially based on the avoidance of sources of error, and consists in the necessity of setting aside certain disadvantages and treating their source; hence it is entitled, when the art is properly appreciated, to be introduced in all such schools.

"Many other grounds can be noticed in favour of this introduction.

"Art and science are learnt in technical schools for practical ends. Practice and a knowledge of drawing are specially demanded on entering the engineering profession. It may even be affirmed, of two equally talented pupils, the best draughtsman will take the first place. Drawing is the centre of gravity for most technical professions, and for this reason alone it ought not to be neglected to promote technical and freehand drawing in every direction. But photography is destined to support these technical studies, as it is also a mode of drawing. Indeed, if it be proposed to draw a complicated machine—as a weaver's wheel—in a few minutes, photography is the only means to do this. The labour, otherwise requiring weeks, is reduced to an affair of a quarter of an hour, and is so perfectly done that all proportions are duly observed, and the projection must be correct from whatever point it is taken, if proper lenses be used.

"If we follow the biography of gifted pupils, we often trace them, aided by Government stipends, going first on distant journeys to study modern and ancient monuments, and to bring home as faithful designs of them as possible. What a severe labour this implies for the architect, amidst a foreign population, in a trying climate, who has to project faithful sketches in a short time amidst countless obstacles! On the other hand, how it is all abridged by photography! What would not the young travelling engineer give to take plans of entire manufactories which he has only a few minutes to view? What would not the highly cultivated philologist give to retain for himself and others, in a durable form, the overpowering impressions of life in the past, which he can only feel as a transient emotion, on the classical ground of Greece and Italy? It is our duty to announce it publicly, that all these wishes have become a possible and a tangible reality through the progress of photography, and that the practice necessary to effect this is easily attained."

Krippendorf omits here an important point, which is the great value of photography to those who are devoted to practical typography, whether it be lithography, book printing, copper-plate printing, printing of paper money, porcelain manufacture or dyeing; for in all these branches the aid of photography is very important. In this connection we refer to the chapter on pyro-photography, heliography, and chromo-photography. In these branches we see photography an auxiliary to the multiplying arts.

Though it has done great things in this connection, we see very few heliographers and photo-lithographers. The ground of this is found simply in the fact that art schools, training lithographers and copper-plate engravers, entirely overlook photography. It is set aside as no art at all by persons who feel themselves artists, yet to whom it would be useful. But in the before-mentioned alliance of photography with lithography and metal-plate printing, good results can only be achieved by the operator being equally skilled in both arts. The author has often witnessed the failure of heliographers, lithographers, and photographers who tried to work by combining the two arts. It is therefore necessary that the schools of art should take the matter in hand, and if so, a new province, hitherto unknown to the lithographer, that of light-printing (see p. 243), must soon become domiciliated in those institutions.

But a knowledge of photography is equally important for painters. Photography copying oil-paintings has taken a magnificent development in our time. Adverse opinions to it are indeed uttered by stiff art critics, such as Thansing, just as idealistic tourists formerly ranted against railways, because travel was thereby robbed of its poetry. These people were right from their point of view, but they could not stop the introduction of railways; and, though travel may have become less poetic through them, these lines have the advantage of giving an opportunity to persons of slender means to make excursions, and thereby to enrich their minds with a knowledge of foreign countries and people, and of improving their health. Photography affords to persons of small income similar advantages in the province of art. Paintings too expensive to be purchased by any save the rich, became slowly and imperfectly known to others by the expensive medium of copper-plate prints. These engravings were also confined to a limited circle. But now photography reproduces with the rapidity of lightning, and with all faithfulness, the latest creations of art, and thus its cheapness makes them accessible to all. The copy is not so artistic as that of the engraver, but it suffices to bring all that is new quickly to the knowledge of all, and in spite—or in consequence rather—of this, the engraving coming after still retains its value.

Moreover, the negatives from oil-paintings require working up by *retouche*, in order to equalize the false effects of colouring. This *retouche* may be very injurious if carried out by unskilled hands. The most suitable hand is that of the painter who painted the original. Good painters have already successfully worked at reproducing negatives from their own originals, and the impressions from plates touched up by the artists themselves must evidently have a much higher value than those emanating from other hands. This presents a fine new field for the artist; but it can only be worked with good results if the art pupils have become familiar in art schools with the technical routine of negative *retouche*, and with taking positive impressions connected with them.

In conclusion we add a few words on the development of the professional photographer.

We have already shown that if portrait and landscape photography are to produce really solid results, they require a knowledge of the principles of art. But hitherto nothing has been done to train photographers artistically. Moreover, photography can only be raised, in an artistic point of view, when art schools render a study of art possible to the photographer. The time must be at hand when all small jealousy directed against photography must fall to the ground. Experience has already found that it is not a rival, but a handmaid to the solidly trained artist.

Photography can be the more readily introduced in schools, as its tuition requires much less time than drawing, and with better results. Four hours a week for six months suffice to train a pupil enough in photography to enable him to go on by himself, even without a knowledge of chemistry.

Schools of science, as well as of art, must attend to this branch, because photography is very useful as an aid to natural science.

This new art gives beautiful illustrations for science and art lectures by the magic lantern. The investigator can by its means give faithful original pictures of the results of his labours (see page 93). Hitherto proper apparatus was wanted; the magic lanterns sold in Germany gave imperfect images. Latterly, R. Talbot in

Berlin, has introduced American magic lanterns, which are best adapted for lectures.

The Woodbury press has been joined to the above, for the illustration of lectures; and the latest improvements in dry plate photography have had the result that dry plates, like *licht-paus* paper, have become articles of trade, making the production of photographs much more easy. Thus one improvement combines with another to make photography what it ought to be—a universal art of writing by light!

AMERICAN APPENDIX.

THERE is an omission in the foregoing work, which, though it may matter little with the European editions, it is worth while to point out in the American edition : the distinguished author has hardly done justice to the science of this country in its contributions to the development of his subject.

Professor Vogel ascribes the first daguerreotypes to Professor Morse; and states that "his coadjutor was Professor Draper." Now, the fact is, that Dr. J. W. Draper, of New York, was the first person who took daguerreotype-portraits from the life, which was in September, 1839. He published an announcement of it in the *London and Edinburgh Philosophical Magazine* of March, 1840, and shortly afterward gave a detailed account of the whole operation in the same journal. Dr. Draper had been already for a considerable time making researches on the chemistry of light, and was not only familiar with the whole subject, but was active in its original exploration. In his laboratory in the New York University, Professor Morse learned the art.

Dr. Draper was also the first to photograph the fixed lines of Fraunhofer in the solar spectrum, and to show that they exist in large numbers beyond the violet space.

He was, besides, the first to decompose carbonic acid in the actual spectrum ; and to prove that it is the yellow light that is efficient in this change, and not the violet rays, as had been formerly supposed.

APPENDIX.

As has been recognized by high foreign authorities, and as is well known, Dr. Draper was the first to devise a method for measuring the intensity of the chemical rays, by which it first became possible to subject them to quantitative investigations.

He was, moreover, the founder of astronomical photography, having taken pictures of the moon, which were exhibited at the New York Lyceum of Natural History, March 23, 1840.

Dr. Draper also established the great principle which is at the basis of spectrum analysis. In a memoir "On the Production of Light by Heat," he showed experimentally that all solid substances, and probably liquids, become incandescent at the same temperature of about 977° Fahr.; that the spectrum of an incandescent solid is continuous, and contains neither bright nor dark fixed lines; that from common temperatures up to 977° Fahr., the rays emitted by a solid are invisible, but at that temperature they impress the eye with the sensation of red; that as the heat of the body continuously rises, other rays are added, increasing in refrangibility as the temperature ascends. The English professor, Roscoe, in the third and last edition of his "Spectrum Analysis," acknowledges that "this law was discovered by Draper;" but the German, Kirchhoff, although he had himself built upon Draper's results, makes no reference to them in his historical sketch of spectrum analysis.

The author of the present work touches upon these various points, but is careless of what Dr. Draper has done—an oversight which is the more marked, as he seems careful to acknowledge the claims of *savants* in the different countries of Europe.

E. L. Y.

NEW YORK, *April*, 1875.

INDEX.

THE END.

THE LITERATURE OF PHOTOGRAPHY

AN ARNO PRESS COLLECTION

Anderson, A. J. **The Artistic Side of Photography in Theory and Practice.** London, 1910

Anderson, Paul L. **The Fine Art of Photography.** Philadelphia and London, 1919

Beck, Otto Walter. **Art Principles in Portrait Photography.** New York, 1907

Bingham, Robert J. **Photogenic Manipulation.** Part I, 9th edition; Part II, 5th edition. London, 1852

Bisbee, A. **The History and Practice of Daguerreotype.** Dayton, Ohio, 1853

Boord, W. Arthur, editor. **Sun Artists** (Original Series). Nos. I-VIII. London, 1891

Burbank, W. H. **Photographic Printing Methods.** 3rd edition. New York, 1891

Burgess, N. G. **The Photograph Manual.** 8th edition. New York, 1863

Coates, James. **Photographing the Invisible.** Chicago and London, 1911

The Collodion Process and the Ferrotype: Three Accounts, 1854-1872. New York, 1973

Croucher, J. H. and Gustave Le Gray. **Plain Directions for Obtaining Photographic Pictures.** Parts I, II, & III. Philadelphia, 1853

The Daguerreotype Process: Three Treatises, 1840-1849. New York, 1973

Delamotte, Philip H. **The Practice of Photography.** 2nd edition. London, 1855

Draper, John William. **Scientific Memoirs.** London, 1878

Emerson, Peter Henry. **Naturalistic Photography for Students of the Art.** 1st edition. London, 1889

*Emerson, Peter Henry. **Naturalistic Photography for Students of the Art.** 3rd edition. *Including* The Death of Naturalistic Photography, London, 1891. New York, 1899

Fenton, Roger. **Roger Fenton, Photographer of the Crimean War.** With an Essay on his Life and Work by Helmut and Alison Gernsheim. London, 1954

Fouque, Victor. **The Truth Concerning the Invention of Photography:** Nicéphore Niépce—His Life, Letters and Works. Translated by Edward Epstean from the original French edition, Paris, 1867. New York, 1935

Fraprie, Frank R. and Walter E. Woodbury. **Photographic Amusements Including Tricks and Unusual or Novel Effects Obtainable with the Camera.** 10th edition. Boston, 1931

Gillies, John Wallace. **Principles of Pictorial Photography.** New York, 1923

Gower, H. D., L. Stanley Jast, & W. W. Topley. **The Camera As Historian.** London, 1916

Guest, Antony. **Art and the Camera.** London, 1907

Harrison, W. Jerome. **A History of Photography Written As a Practical Guide and an Introduction to Its Latest Developments.** New York, 1887

Hartmann, Sadakichi (Sidney Allan). **Composition in Portraiture.** New York, 1909

Hartmann, Sadakichi (Sidney Allan). **Landscape and Figure Composition.** New York, 1910

Hepworth, T. C. **Evening Work for Amateur Photographers.** London, 1890

*Hicks, Wilson. **Words and Pictures.** New York, 1952

Hill, Levi L. and W. McCartey, Jr. **A Treatise on Daguerreotype.** Parts I, II, III, & IV. Lexington, N.Y., 1850

Humphrey, S. D. **American Hand Book of the Daguerreotype.** 5th edition. New York, 1858

Hunt, Robert. **A Manual of Photography.** 3rd edition. London, 1853

Hunt, Robert. **Researches on Light.** London, 1844

Jones, Bernard E., editor. **Cassell's Cyclopaedia of Photography.** London, 1911

Lerebours, N. P. **A Treatise on Photography.** London, 1843

Litchfield, R. B. **Tom Wedgwood, The First Photographer.** London, 1903

Maclean, Hector. **Photography for Artists.** London, 1896

Martin, Paul. **Victorian Snapshots.** London, 1939

Mortensen, William. **Monsters and Madonnas.** San Francisco, 1936

***Nonsilver Printing Processes:** Four Selections, 1886-1927. New York, 1973

Ourdan, J. P. **The Art of Retouching by Burrows & Colton.** Revised by the author. 1st American edition. New York, 1880

Potonniée, Georges. **The History of the Discovery of Photography.** New York, 1936

Price, [William] Lake. **A Manual of Photographic Manipulation.** 2nd edition. London, 1868

Pritchard, H. Baden. **About Photography and Photographers.** New York, 1883

Pritchard, H. Baden. **The Photographic Studios of Europe.** London, 1882

Robinson, H[enry] P[each] and Capt. [W. de W.] Abney. **The Art and Practice of Silver Printing.** The American edition. New York, 1881

Robinson, H[enry] P[each]. **The Elements of a Pictorial Photograph.** Bradford, 1898

Robinson, H[enry] P[each]. **Letters on Landscape Photography.** New York, 1888

Robinson, H[enry] P[each]. **Picture-Making by Photography.** 5th edition. London, 1897

Robinson, H[enry] P[each]. **The Studio, and What to Do in It.** London, 1891

Rodgers, H. J. **Twenty-three Years under a Sky-light,** or Life and Experiences of a Photographer. Hartford, Conn., 1872

Roh, Franz and Jan Tschichold, editors. **Foto-auge, Oeil et Photo, Photo-eye.** 76 Photos of the Period. Stuttgart, Ger., 1929

Ryder, James F. **Voigtländer and I:** In Pursuit of Shadow Catching. Cleveland, 1902

Society for Promoting Christian Knowledge. **The Wonders of Light and Shadow.** London, 1851

Sparling, W. **Theory and Practice of the Photographic Art.** London, 1856

Tissandier, Gaston. **A History and Handbook of Photography.** Edited by J. Thomson. 2nd edition. London, 1878

University of Pennsylvania. **Animal Locomotion. The Muybridge Work at the University of Pennsylvania.** Philadelphia, 1888

Vitray, Laura, John Mills, Jr., and Roscoe Ellard. **Pictorial Journalism.** New York and London, 1939

Vogel, Hermann. **The Chemistry of Light and Photography.** New York, 1875

Wall, A. H. **Artistic Landscape Photography.** London, [1896]

Wall, Alfred H. **A Manual of Artistic Colouring, As Applied to Photographs.** London, 1861

Werge, John. **The Evolution of Photography.** London, 1890

Wilson, Edward L. **The American Carbon Manual.** New York, 1868

Wilson, Edward L. **Wilson's Photographics.** New York, 1881

All of the books in the collection are clothbound. An asterisk indicates that the book is also available paperbound.